T0192256

SpringerBriefs in Applied Sciences and Technology

SpringerBriefs present concise summaries of cutting-edge research and practical applications across a wide spectrum of fields. Featuring compact volumes of 50 to 125 pages, the series covers a range of content from professional to academic.

Typical publications can be:

- A timely report of state-of-the art methods
- An introduction to or a manual for the application of mathematical or computer techniques
- A bridge between new research results, as published in journal articles
- A snapshot of a hot or emerging topic
- An in-depth case study
- A presentation of core concepts that students must understand in order to make independent contributions

SpringerBriefs are characterized by fast, global electronic dissemination, standard publishing contracts, standardized manuscript preparation and formatting guidelines, and expedited production schedules.

On the one hand, **SpringerBriefs in Applied Sciences and Technology** are devoted to the publication of fundamentals and applications within the different classical engineering disciplines as well as in interdisciplinary fields that recently emerged between these areas. On the other hand, as the boundary separating fundamental research and applied technology is more and more dissolving, this series is particularly open to trans-disciplinary topics between fundamental science and engineering.

Indexed by EI-Compendex, SCOPUS and Springerlink.

More information about this series at https://link.springer.com/bookseries/8884

Benjamin Aziz

Formal Analysis by Abstract Interpretation

Case Studies in Modern Protocols

Benjamin Aziz (iD)
School of Computing
University of Portsmouth
Portsmouth, UK

ISSN 2191-530X   ISSN 2191-5318   (electronic)
SpringerBriefs in Applied Sciences and Technology
ISBN 978-3-030-91152-2   ISBN 978-3-030-91153-9   (eBook)
https://doi.org/10.1007/978-3-030-91153-9

This Springer imprint is published by the registered company Springer Nature Switzerland AG
The registered company address is: Gewerbestrasse 11, 6330 Cham, Switzerland

To my loving family

Preface

The book outlines how formal static analysis techniques can be defined and applied to modern protocols related to new paradigms such as the IoT, Industry 4.0 and Smart Energy. The application of such robust techniques can lead to better understanding of modern-day protocols and systems, both within the context of classical requirements, such as functional security, and the more challenging properties related to the presence of vulnerabilities at levels of specification and design. The book challenges current mainstream thinking that large-scale systems are simply infeasible and non-suitable case studies for formal analysis techniques, and goes to demonstrate that such systems can benefit from the application of such techniques.

The book gives us a static analysis framework with a focus on capturing name substitutions occurring as a result of processes communicating with one another. First, a non-standard denotational semantics for the formal language is constructed based on domain theory. The semantics is shown to be sound with respect to transitions in the standard structural operational semantics. Finally, the non-standard semantics is abstracted to operate over finite domains so as to ensure the termination of the static analysis. The safety of the abstract semantics is proven with respect to the non-standard semantics. This framework demonstrates that the choice of the name-substitution property is a fundamental one, in that it can be used to understand a wide range of system and protocol properties ranging from security and safety-related ones to testing coverage to mutation-based properties. As such, the book lays down a simple but effective method for the static analysis of systems and protocols that go beyond simple toy examples into the realm of complex systems.

One of the main motivations in writing this book was to give the reader the benefit of understanding how a highly theoretical area in computer science can have applications in complex industrial settings, two universes that are often hard to combine. More specifically, it is envisaged that the readers will first gain knowledge and experience of the abstract interpretation method of formal analysis, and how it can be varied, to express various properties, and applied for different purposes, for example, in testing, mutation and semantic ambiguity resolution contexts. Second, they will be challenged to think of how such method can be used in analysing modern protocols, in other words, protocols of significance of complexity as opposed to the majority of current literature that applies only to toy-style protocols. The expectation is that the book will encourage readers, in particular those working in research, to think and reflect on new directions for future research benefiting from the approach discussed throughout.

The book is composed of five chapters along with the introduction (Chap. 1) and the conclusion (Chap. 7). From these five chapters, the first two (Chaps. 2 and 3) define the formal language and its denotational semantics, both concrete and abstract, and hence are an essential reading in order to understand any of the three case studies presented in the following three chapters (Chaps. 4–6). The case studies themselves can be read individually, in any order, and according to the needs of the reader, as they are not inter-related.

The primary audience of this book is the academic community (postgraduate and researchers), particularly at the intersection of formal methods, IoT systems, networks and program analysis areas. The book is also suitable for the industrial research community, particularly those working with the IoT and communication networks research. Finally, the book is also expected to be useful to anyone interested in the standardisation of protocols, particularly, IoT and Industry 4.0 protocol standards.

Special thanks to my Ph.D. supervisor, Dr. Geoff Hamilton, whose comments and feedback over the years were inspirational in defining the static analysis framework.

Basingstoke, UK   Benjamin Aziz
September 2021

Contents

Acronyms

BNF	Backus–Naur Form
CA	Certification Authority
CCS	Calculus of Communicating Systems
CFA	Control Flow Analysis
CoAP	Constrained Application Protocol
CoSeC	Compositional Security Checker
CPO	Complete Partial Order
CPS	Cyber-Physical System
CS	Charging Station
CSMS	Charging Station Management System
CSP	Communicating Sequential Processes
EV	Electric Vehicle
EVSE	Electric Vehicle Supply Equipment
FTP	File Transfer Protocol
HTML	HyperText Markup Language
IIoT	Industrial Internet of Things
IoT	Internet of Things
IP	Internet Protocol
IPC	Institute for Printed Circuits
IPC-CFX	Institute for Printed Circuits Connected Factory Exchange
LAN	Local Area Network
LF	(Edinburgh) Logical Framework
LMAC	Lightweight Medium Access
MiM	Man-in-the-Middle
ML	Meta-Language
MMC	Mobility Model Checker
MOM	Message Oriented Middleware
MQTT	Message Queuing Telemetry Transport
OASIS	Organization for the Advancement of Structured Information Standards
OCA	Open Charge Alliance
OCPP	Open Charge Point Protocol

OPC	Open Platform Communications
PCB	Printed Circuit Board
PKI	Public Key Infrastructure
PPMP	Production Performance Management Protocol
QoS	Quality of Service
RD	Reaching Definitions
RFID	Radio Frequency Identification
RMI	Remote Method Invocation
RPC	Remote Procedure Call
SMV	Symbolic Model Checking
SSL	Secure Sockets Layer
TCP	Transmission Control Protocol
UML	Unified Modelling Language
VDL	Vienna Definition Language

Chapter 1
Introduction

1.1 Background

In order to understand the approach we adopt in this book in the analysis of the three examples of modern complex systems, we give here some overview of the three main areas that form the basis for this approach: Calculi describing mobility, denotational semantics and static program analysis.

1.1.1 Calculi Describing Mobility

The word *mobile* in the real world is normally used to describe the state of any object, location, condition, etc. that is moving with respect to some reference. In the context of computing systems, mobility may refer to the movement of communication channels, code, or whole computing environments. For example, HTML links can be created, sent to other entities and destroyed later. References to objects in object-oriented programming are created and passed around as capabilities of communication. In Java, Applet code embedded in Web pages can be downloaded and executed dynamically at runtime by the host machine. The movement of "intelligent" mobile agents constitutes a form of code and state mobility. Finally, the emerging component-based technologies and pervasive/ubiquitous computing are interesting examples of mobile computing environments.

From a process algebraic point of view, there are two widely accepted definitions of mobility [1]:

1. *Link-based mobility.*
   The concept of link-based mobility states that the movement of a process among other processes can in fact be described as the proliferation, change and extinction of the communication channels linking that process to the rest. In other words, this corresponds to the movement of links in the virtual space of linked processes. The basic notion of communication here is that of process interaction, where processes

© The Author(s), under exclusive license to Springer Nature Switzerland AG 2022
B. Aziz, *Formal Analysis by Abstract Interpretation*, SpringerBriefs in Applied
Sciences and Technology, https://doi.org/10.1007/978-3-030-91153-9_1

interact to exchange links. This induces the behaviour of message passing. The resulting mobility is expressively powerful and can be used to encode the higher order notion of process-based mobility, in which whole processes (not just links) can move in the virtual space of linked processes. The π-calculus [1–3] is among the most authoritative models that embrace the link-based definition of mobility.

2. *Location-based mobility.*

   Unlike the link-based definition of mobility, which describes a virtual movement of links, the location-based definition relies on the physical aspect of mobility. According to this definition, mobility is the movement of processes in the space of locations. For example, the movement of a laptop from one local area network to another is viewed as the movement of a computing process (the laptop) from one parent location (first LAN) to another (second LAN). The best example of a formalism that adopts the concept of location-based mobility is the Mobile Ambient calculus [4], which is mainly influenced by Internet programming and the presence of administrative domains that divide wide area networks in general.

By and large, the modelling of mobility in computing systems has benefited a good deal from the body of theory that was developed earlier for the modelling of static concurrent and distributed systems. Formalisms such as Petri nets [5], CSP [6] and CCS [7] provided the necessary mature ground for the arrival of the first substantial theory of mobility, the π-calculus. The development of the π-calculus was directly inspired by the calculus of [8], where label-passing was added to the theory of CCS to model the dynamic configuration of networks. In the π-calculus, the theory was further simplified by adopting the unique notion of a *name* (hence identifying variables and constants). Names refer to channels of communication and can be communicated over other names. The concept of mobility is grasped by allowing processes to exchange names of channels, and hence modify the network configuration dynamically.

The π-calculus is a highly expressive language that is also capable of encoding statically distributed systems and functional programming (λ-calculus). The language is characterised as being directly executable and has formed the basis for other programming languages like Piccola [9], Join [10], Pict [11] and Nomadic Pict [12]. It has also provided the basis for many extended models that are concerned with different aspects and properties of mobile systems. The extension we are interested in is the spi calculus [13], which extends the language of the π-calculus by the addition of cryptographic primitives like encryption/decryption and digital signing/verification. Reasoning about the different properties of cryptographic protocols is based on a theory of testing equivalence. Properties like privacy and authentication are defined as equivalences in the presence of intruder processes.

The popularity of the π-calculus led to the development of the Mobile Ambient calculus that adopts a different approach in the modelling of mobility. Unlike the location-transparent π-calculus, the Mobile Ambients adopts the notion of an *ambient* as its main idea, and it is most suitable for the modelling of computing agents that can move from one location to another. Ambients are bounded places where computations can take place. The boundary of an ambient is significant to its movement, since it determines exactly what entity will move. It is also significant from the security perspective as it acts as an access control mechanism determining what boundaries can or cannot be crossed. Examples of ambients include a virtual address space, a laptop, a Unix file system and a single Java object. Furthermore, ambients may be nested within other ambients.

Among other formalisms, which adopt definitions of mobility that overlap definitions (1) and (2) above and that have been shown to be interesting, is the seal calculus [14]. *Seals* are named locations that are passed around and can have portals opened for remote communications occurring with the parent and child seals, while allowing for local communications to take place directly within the same seal. The seal calculus extends the polyadic π-calculus [15], and in comparison to the Mobile Ambient calculus, it adopts an *objective* movement of locations; movement is initiated by the environment surrounding a seal. The Mobile Ambient calculus, on the other hand, adopts *subjective* mobility allowing ambients to initiate the movement.

The seal calculus elegantly models Internet programming. It adheres to a number of principles that are reminiscent of Internet-like programming. These include the distinction between remote and local communications, locations, restricted connectivity and access control. In particular, the modelling of security notions that rely on locations and scoping, like the perfect firewall equation, is straightforward.

One of the interesting issues currently debated within the area of mobility formalisms and security is whether there is a need for combining location-based mobility languages, like the Mobile Ambients and seal calculi, with cryptographic primitives in a similar manner to the spi calculus. The analogy between the concepts of a mobile location and a ciphertext encrypted with a symmetric key is prevalent. However, it is less obvious when dealing with asymmetric-key cryptography and operations like hashing. Nonetheless, the modelling of mechanisms, like remote communications (e.g. RPC and RMI), could well benefit from such a combination.

In our book, we shall adopt a formal language based on the π-calculus enhanced with a special construct of *timed inputs*, where input actions could have a limited time duration within which they are active.

1.1.2 Denotational Semantics

The denotational (also known as the mathematical) approach to the definition of the semantics of programming languages was initiated by Christopher Strachey and instrumented by Dana Scott in the late 1960s and early 1970s [16–21]. The idea suggested by Scott and Strachey was to develop a mathematical framework within which the formal semantics of programming languages could be specified without the traditional implementation-dependent problems associated with operational semantics and that would rely on the rigour that mathematics offers. Indeed, this framework later became an inspiration for computer language designers and implementers.

The basis of any denotational model is that syntactic phrases are realisations of abstract mathematical objects. For example, in a calculator device, strings of digits are perceived as abstract numerical ideas regardless of the format in which those digits are presented on the screen. Another example is the functional view that programs stand for mathematical functions, and the execution of a program, given some input data, resembles the application of a function to its parameters.

In general, for any syntactically correct program, there exists a mathematical object known as the *denotation* of that program that expresses the meaning of the program in a clear and non-circular (i.e. needless of further definitions) manner. Based on this, one may divide a denotational model into three components:

1. *Syntactic domains.* These are the collection of entities that constitute the syntactical representation of the language constructs and whose meaning is sought. Syntactic domains include digits, numerals, expressions, instructions, phrases and programs. The syntactic domains we adopt here are *abstract* [22], as opposed to the concrete syntactic domains, which normally introduce unnecessary syntactic sugar useful only for the parsing of programs and do not contribute to their meaning. Furthermore, the syntactic notation we adopt is a version of the Backus–Naur Form (BNF) [23], which describes context-free grammars. Classical readings on the subject of syntax include [24, 25].
2. *Semantic domains.* These are collections of mathematical objects (denotations) that convey the meaning of the syntactical entities. Elements of these objects usually have some structure, like complete partial orders (CPOs), lattices or domains, whose algebra is determined by domain theory. The semantic elements are usually classified as either primitive or complex elements, where the bottom element in these domains often denotes the undefined program. Primitive elements constitute the *atomic* semantic elements, whereas complex elements are necessary to convey more sophisticated ideas that can be decomposed back to the primitive elements. For example, it is common in the theory of the π-calculus to interpret parallelism in terms of the simpler notions of input/output and non-determinism.
3. *Semantic functions.* These are special functions that map programs, phrases, etc. from their syntactic domains to their denotations in the semantic domains. Often, there are certain requirements that need to be satisfied by semantic functions, for example, being monotonic and continuous.

In general, denotational semantics has to respect the principle of *compositionality*. Compositionality states that the meaning of a program can be defined in terms of the meanings of its subprograms. A discussion on the principle of compositionality in the definition of language semantics can be found in [26]. Also, the well-known *Pisa Notes* [27] on domains compiled by Gordon Plotkin are a traditional reading in the subject of domain theory and its application to semantics.

One of the main advantages that the denotational semantics approach introduced is that the behaviour of any program could be determined directly through the mathematics of domain theory without the need to execute that program, and consequently, without the need to design any language compilers or interpreters. As a result, program verification and comparison becomes an easier task compared to other approaches to the definition of language semantics.

The fact that the denotational approach relies, to a large degree, on domain theory means that several powerful mathematical tools become available for reasoning about program semantics. Mainly, concepts like CPOs, continuous functions and least fixed points are easier to express and implement. For example, it is often the case that demonstrating the termination of static analysis is dependent on the evaluation of the least fixed points.

Finally, the idea of using semantic functions to denote the meaning of programs created a close relationship between the theory of denotational semantics and the functional programming paradigm. The task of implementing denotational interpreters using functional languages becomes fairly straightforward. Any static analysis that, in turn, is based on the denotational semantics of the language can also be directly implemented as a higher level abstraction (i.e. semantics-directed). The implementations of fixed points are also standard in functional languages.

1.1.3 Static Program Analysis

Quite often, it is desirable to predict in advance the set of values resulting from programs and verify certain properties regarding their runtime behaviour. For this purpose, the area of static program analysis offers compile-time computable techniques that can be used to safely approximate the properties and values of programs without the need to execute them directly on computers. The functionality covered by static analysers is wide and ranges from simple syntactic verifications that can be used in program transformations to complex runtime properties related to issues of security and optimisation. In general, many approaches exist for building static analysers. Here, we distinguish four main approaches similar to [28]: Data flow analysis, control flow analysis, abstract interpretation and type systems. In the rest of the book, we shall focus on the abstract interpretation approach.

1.1.3.1 Data Flow Analysis

Motivated by the aim of producing smaller and faster programs, the main application area of the data flow analysis approach has always been the performance optimisation of program code generated by compilers. Other important applications that benefit from data flow analysis include program testing, validation, verification, parallelisation and vectorisation. Classical data flow analyses include reaching definitions, available expressions and live variables analyses [28].

A data flow analysis is primarily designed to gather information about the use and definition of data in a program as well as the dependencies between the different sets of data. To achieve this, a program is often seen as a graph, where nodes represent blocks of the program and edges represent flows between those blocks. For example, the following program computes the factorial of a number n:

Fig. 1.1 The flow graph of
the factorial program

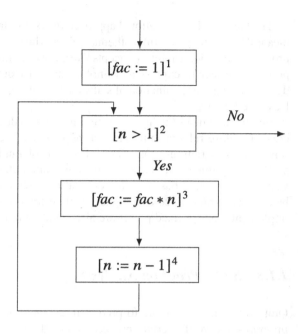

$$F \stackrel{\text{def}}{=} [fac := 1]^1; \texttt{while } [n > 1]^2 \texttt{ do } ([fac := fac * n]^3; [n := n - 1]^4)$$

This program can be represented by the graph of Fig. 1.1, and can also be expressed by the function $flow = \{(1, 2), (2, 3), (3, 4), (4, 2)\}$.

Sets of equations or constraints are then constructed that relate the entry and exit information of each program node and among the different nodes. The least solution of these equations represents the result of the analysis. In the above example of the factorial program, one may construct a *reaching definition analysis* by defining the following set of equations [28]:

$$Rout(1) = (Rin(1)\backslash\{(fac, l)\}) \cup \{(fac, 1)\}$$
$$Rout(2) = Rin(2)$$
$$Rout(3) = (Rin(3)\backslash\{(fac, l)\}) \cup \{fac, 3\}$$
$$Rout(4) = (Rin(4)\backslash\{(n, l)\}) \cup \{(n, 4)\}$$
$$Rin(2) = Rout(1) \cup Rout(4)$$
$$Rin(3) = Rout(2)$$
$$Rin(4) = Rout(3)$$

where l is any label, $l \in \{1, 2, 3, 4\}$. *Rout*(l) here relates information at the exit point of a node, l, in terms of the inputs to that node, and *Rin*(l) relates information about the entry point of node, l, in terms of the outputs from that point. The restriction, $Rin(l')\backslash\{(x, l)\}$, and the union, $Rin(l') \cup \{(x, l')\}$, refer to the change that occurs

l	$Rin(l)$	$Rout(l)$
1	$(n, ?), (fac, ?)$	$(n, ?), (fac, 1)$
2	$(n, ?), (n, 4), (fac, 1), (fac, 3)$	$(n, ?), (n, 4), (fac, 1), (fac, 3)$
3	$(n, ?), (n, 4), (fac, 1), (fac, 3)$	$(n, ?), (n, 4), (fac, 3)$
4	$(n, ?), (n, 4), (fac, 3)$	$(n, 4), (fac, 3)$

Fig. 1.2 The solution of the reaching definition analysis for the factorial program

in the value of variable x at program node l', where the previous value of x was obtained at node l. On the other hand, $Rin(l)$ reflects the information collected from nodes whose outputs flow into the input of node l. Hence, $Rout(l')$ is added to the value of $Rin(l)$ whenever control flow is possible from node l' to node l.

In a reaching definition analysis, the information collected about a program point represents the assignments that can reach that point, i.e. assignments not overwritten by the time the point is reached. Hence, solving the above equations will result in the least solution of Fig. 1.2 (the question mark "?" symbol denotes unknown value).

Data flow analysis is a mature research area, where frameworks have been developed to classify and solve different classes of data flow problems. These problems can be described using the formalisms of the different frameworks, and a solution algorithm is then selected. The earliest framework for data flow analysis can be ascribed to Kildall [29], who was also the first to use semi-lattices in such analyses. A survey can be found in [30], which includes the rapid, continuous, distributive, monotone and k-bound frameworks.

The most popular framework for data flow analysis is the monotone framework due to [31]. In this framework, data flow problems are defined through lattices (either complete or semi-lattices) of values with a meet (join) operator, often called the *property space*, and a family of transfer functions defined over those lattices. To create an instance of the framework, a (directed) flow graph is required, where the transfer functions are bound to the graph nodes using a function map. The association of a function to a particular node means that all the information gathered at the node is mapped to lattice values. This also allows for data flow problems to be phrased as equations that encode the information at each node in terms of the information at the predecessor (successor) nodes and the local transfer functions. The solution to a problem is then obtained by computing a fixed point (least or greatest) of the equation system, where a lattice value is associated with each node of the flow graph. An interesting treatment of the monotone framework for concurrent systems is given in [32].

The only application of the data flow paradigm to the analysis of the π-calculus has been given in [33], where the issue of causality among processes and true concurrency is examined. However, as we are unaware of any security-related data flow analyses, we shall not discuss this paradigm any further.

1.1.3.2   Control Flow Analysis

A control flow analysis is concerned with answering the following question: Given a particular point in a program, what is the set of subprograms, functions, commands, etc. that can be reached from that point? In other words, a control flow analysis attempts to record information about the different execution paths in the program. This information is then used to conduct program optimisation and transformation as well as determine runtime properties. Despite the fact that the early control flow analysis techniques were developed mainly for the functional programming paradigm, control flow analyses were later utilised in other paradigms as well, like the object-oriented, concurrent and logic programming paradigms.

Often, a control flow analysis is expressed as a constraint-based analysis, and its solution relates the different points of the control flow of the basic program blocks. Taking a functional program as an example, the set of constraints may be classified into three main classes, assuming a finite set of program labels. The first two classes relate function abstractions and variable values to their corresponding labels, respectively. The third class expresses which formal parameters of functions are bound to which actual parameters in function applications, and whether the results returned from functions can be returned by the applications in which they appear.

For example, consider the following functional program [28]:

$$[[\texttt{fn x => } [\texttt{x}]^1]^2 \, [\texttt{fn y => } [\texttt{y}]^3]^4]^5$$

If we assume functions $C(l)$ and $label(x)$, where the former indicates the set of expressions that l may evaluate to, and the latter the values that variable x may be bound to, then the following classes of constraints may be stated for the above program. The first class relates values of function abstractions to their labels:

$$\{\texttt{fn x => } [\texttt{x}]^1\} \subseteq C(2)$$
$$\{\texttt{fn y => } [\texttt{y}]^3\} \subseteq C(4)$$

The second class of constraints relates values of variables to their labels:

$$label(x) \subseteq C(1)$$
$$label(y) \subseteq C(3)$$

The final class expresses information about function applications, where the following conditional constraints are introduced (the inclusion of functions fn x => x and fn y => y provides for the possibility that both functions may be applied):

$$\{\texttt{fn x => [x]}^1\} \subseteq C(2) \Rightarrow C(4) \subseteq \texttt{label(x)}$$
$$\{\texttt{fn x => [x]}^1\} \subseteq C(2) \Rightarrow C(1) \subseteq C(5)$$
$$\{\texttt{fn y => [y]}^3\} \subseteq C(2) \Rightarrow C(4) \subseteq \texttt{label(y)}$$
$$\{\texttt{fn y => [y]}^3\} \subseteq C(2) \Rightarrow C(3) \subseteq C(5)$$

The least solution to the above equations is given as follows:

$$C(1) = \{\texttt{fn y => [y]}^3\}$$
$$C(2) = \{\texttt{fn x => [x]}^1\}$$
$$C(3) = \{\,\}$$
$$C(4) = \{\texttt{fn y => [y]}^3\}$$
$$C(5) = \{\texttt{fn y => [y]}^3\}$$
$$\texttt{label(x)} = \{\texttt{fn y => [y]}^3\}$$
$$\texttt{label(y)} = \{\,\}$$

This solution reveals that the function (fn y => y) is never applied in the program (since we have that label(y)={}), and the program may only evaluate to the function, fn y => y (since we have that $C(5) = \{\texttt{fn y => [y]}^3\}$). Hence, the last two conditional constraints have false left sides leading to false right sides.

To achieve a more precise control flow analysis, the concept of k-CFA analysis (CFA stands for *Control Flow Analysis*) was developed in [34], where k stands for the level of context information taken into account in the analysis. Hence, a 0-CFA denotes a context-insensitive or *monovariant* analysis. On the other hand, when $k > 0$, the analysis is described as context-sensitive or *polyvariant*. The presence of dynamic context information allows for the different instances of variables and program blocks to be distinguished and therefore, arrive at more precise results for the analysis. Several variations of the k-CFA analysis exist, for example, the uniform k-CFA, polynomial k-CFA and the Cartesian product algorithm.

Control flow analysis has been combined with other approaches, like data flow analysis, abstract interpretation and flow graphs, to achieve better quality, understanding and presentation of the final results. Data flow information may be included in the final set of results obtained from the control flow analysis, which then results in a better quality of control flow information. The use of abstract interpretation techniques is necessary for tuning the complexity of the analysis versus its precision. As with the data flow analysis, flow graphs are a handy tool to visualise results and gain a better understanding of the flow of control among different program blocks.

1.1.3.3 Abstract Interpretation

An abstract interpretation allows for programs to be analysed by running their specifications over finite approximated semantic domains that are less precise than the concrete semantic domains, but that are characterised as being safe computable abstractions of the concrete domains. In fact, the original work by [35] has developed from being a specific framework for imperative languages to a general framework offering solutions on the design of static analyses for different programming paradigms. It even became closely linked with other approaches like data flow analysis, control flow analysis and type systems.

The first step in designing an abstract interpretation is to determine whether the standard semantics of the language is sufficiently rich to be able to capture the property under consideration. If not, a non-standard semantics is designed as an extension or modification of the standard semantics to capture the property of interest. The resulting non-standard semantics is sometimes proven to be *correct* with respect to the standard semantics. Different approaches exist for proving the existence of a correctness relation. The approach we follow in our framework is to show that for all the non-standard semantics elements, the standard semantic component can be extracted from these elements.

More formally, assume P is a program, $(\!|P|\!) = v \in V$ is the standard semantic evaluation and $[\![P]\!] = l \in L$ is the non-standard semantic evaluation, then the correctness relation R is formalised as a compositional function:

$$\forall P : \ R([\![P]\!]) = (\!|P|\!)$$
$$\forall v_1, v_2 \in V, l_1, l_2 \in L : \ R(l_1) = v_1 \ \wedge \ R(l_2) = v_2 \ \Rightarrow \ R(l_1 \star l_2) = v_1 \star v_2$$

where $\star$ is a composition operation that constructs the complex meaning, $l_1 \star l_2$, from the primitive meanings l_1 and l_2. Often proving the existence of R is largely dependent on the amount of information about the standard semantics retained in the non-standard semantics.

Since the concrete semantics (standard or non-standard) may operate over infinite domains, the computation of this semantics is not guaranteed to terminate, even with the use of the least fixed points. Therefore, a suitable approximation (abstraction) is required to keep the semantic domain finite. This abstraction is shown to be safe with respect to the concrete semantics by proving that a safety relation is preserved across the abstract semantic values. This relation expresses the fact that every concrete computation maps to a corresponding abstract computation, although the latter, of course, is less precise.

Hence, if we assume that $[\![P]\!]^{\sharp} = l^{\sharp} \in L^{\sharp}$ is the abstract semantic function, then the safety requirement can be formulated as a relation, S, defined as follows:

$$\forall P, [\![P]\!] = l, [\![P]\!]^{\sharp} = l^{\sharp} : (l, l^{\sharp}) \in S$$
$$\forall l_1, l_2 \in L, l_1^{\sharp}, l_2^{\sharp} \in L^{\sharp} : (l_1, l_1^{\sharp}) \ \wedge \ (l_2, l_2^{\sharp}) \in S \ \Rightarrow \ (l_1 \star l_2, l_1^{\sharp} \star l_2^{\sharp}) \in S$$

The definition of the safety relation S is highly dependent on the semantic domains and the choice of the abstraction adopted.

In the theory of abstract interpretation, the set of abstract semantic values, $L^{\sharp}$, is interesting because it is augmented with some ordering relation, $\sqsubseteq$, which results

in $L^\sharp$ having some structure, like a complete lattice. Moreover, we can impose the
following implications between the complete lattice, $L^\sharp$, and the safety relation, S:

$$\forall l \in L, l_1^\sharp, l_2^\sharp \in L^\sharp : (l, l_1^\sharp) \in S \wedge l_1^\sharp \sqsubseteq l_2^\sharp \;\Rightarrow\; (l, l_2^\sharp) \in S$$
$$\forall l \in L, l^\sharp \in L'^\sharp, L'^\sharp \subseteq L^\sharp : (l, l^\sharp) \in S \;\Rightarrow\; (l, \sqcap L'^\sharp) \in S$$

The first implication states that if an abstract value, $l_1^\sharp$, is safe, then a larger value,
$l_2^\sharp$, will also be safe and, therefore, the smaller value, $l_1^\sharp$, would constitute a more
precise or better solution. The second implication states that for a set of safe abstract
values, their greatest lower bound is also a safe value. In other words, for a concrete
semantic value, l, there is always a smallest abstract value, $\sqcap L'^\sharp$, among a sub-
lattice, $L'^\sharp \subseteq L^\sharp$, of safe values, that is itself safe with respect to l. Practically, this
has the effect that an abstract interpretation needs to be performed only once for a
program to obtain the best solution.

An alternative approach to the safety proof involves the introduction of a pair
of abstraction/concretisation relations that are shown to form a *Galois connection* between the concrete and abstract domains. Hence, a tuple, $(L, \alpha, \gamma, L^\sharp)$, is
a Galois connection between the two complete lattices, $(L, \sqsubseteq, \sqcap, \sqcup, \bot, \top)$ and
$(L^\sharp, \sqsubseteq, \sqcap, \sqcup, \bot, \top)$, where $\alpha : L \to L^\sharp$ and $\gamma : L^\sharp \to L$ are monotone functions,
if the tuple satisfies the requirement that

$$\gamma \circ \alpha \;\sqsupseteq\; \lambda l.l$$
$$\alpha \circ \gamma \;\sqsubseteq\; \lambda m.m$$

which express the safety of the abstract semantics at the cost of losing precision. On
certain occasions, *widening* and *narrowing* techniques may also be used in combination with Galois connections to further approximate fixed points.

The main application of abstract analysers has been in combination with language compilers that could be used in program optimisation and to prove that the
program is safe with respect to a certain security policy. Examples of popular non-
security-related implementations of abstract interpretation-based analyses include
the strictness analysis, sharing analysis, ground substitutions analysis in logic programs and approximations of n-dimensional vector spaces over integers and rational
numbers.

1.1.3.4 Type Systems

Type systems are widely used in programming languages to avoid unwanted
behaviour of their programs during runtime. Some of this behaviour could be cru-
cial to security violations, like private information leaks and restricted address space
accesses. In general, a *type* is regarded as holding information that is true about the
program entity it types. Moreover, if one thinks of a type as being a collection of
values, then a subset of that collection constitutes a *sub-type*. Sub-typing expresses

$(assign)$ $[x := a]^l : RD \rightarrow ((RD \setminus \{(x, l')\}) \cup \{(x, l)\})$

$(skip)$ $[\texttt{skip}]^l : RD \rightarrow RD$

(seq) $$\dfrac{S_1 : RD_1 \rightarrow RD_2 \ \ S_2 : RD_2 \rightarrow RD_3}{S_1; S_2 : RD_1 \rightarrow RD_3}$$

(if) $$\dfrac{S_1 : RD_1 \rightarrow RD_2 \ \ S_2 : RD_1 \rightarrow RD_2}{\texttt{if } [b]^l \texttt{ then } S_1 \texttt{ else } S_2 : RD_1 \rightarrow RD_2}$$

$(while)$ $$\dfrac{S : RD \rightarrow RD}{\texttt{while } [b]^l \texttt{ do } S : RD \rightarrow RD}$$

(sub) $$\dfrac{S : RD_2 \rightarrow RD_3}{S : RD_1 \rightarrow RD_4} \ \text{ if } RD_1 \subseteq RD_2 \text{ and } RD_3 \subseteq RD_4$$

Fig. 1.3 Types for the reaching definition analysis

an ordering relation among the different types. It also may be thought of as giving more refined information about the sub-typed entity.

One may also define a *principal type* as the most general type of an expression. For example, a lambda abstraction, $\lambda x.x$, has the principal type, $a \rightarrow a$, where a could be instantiated with any type. Principal types of language expressions are often computed from the types of their sub-expressions using Robinson's unification algorithms [36].

A *typing environment* is used to map the different program entities (statements, expressions, constants, variables, etc.) to their types. Such an environment may be constructed manually according to a set of axioms and rules. Consider the following imperative language:

$$S, S_1, S_2 := [x := a]^l$$
$$\qquad | \ \ [\texttt{skip}]^l$$
$$\qquad | \ \ S_1; S_2$$
$$\qquad | \ \ \texttt{if } [b]^l \texttt{ then } S_1 \texttt{ else } S_2$$
$$\qquad | \ \ \texttt{while } [b]^l \texttt{ do } S$$

where $l, l' \in \{1, 2, 3, 4\}$. Then each statement may be considered as typed by the type $\Sigma \rightarrow \Sigma$, where Σ is the type of the state of the program. This type could denote the result of a reaching definition analysis, RD. Based on this interpretation of Σ, one may introduce the axioms and rules of Fig. 1.3 [28].

These axioms and rules are explained as follows. Axiom (*assign*) states that any previous assignments carried out on the variable x are removed from the final state RD, and x is declared as having the value assigned at current statement marked l. Axiom (*skip*) does not alter RD and rule (*seq*) composes the type of a sequential statement, $S_1; S_2$, from the types of the composed statements, S_1 and S_2. Rule (*if*) deals with the conditional statement, which is given the type of either branch. Rule (*while*) assigns to a while statement the type of its body. Finally, rule (*sub*) deals with the sub-typing of statements.

Given the factorial program of Sect. 1.1.3.1:

$$F \stackrel{\text{def}}{=} [fac := 1]^1; \text{ while } [n > 1]^2 \text{ do } ([fac := fac * n]^3; [n := n - 1]^4)$$

It is possible to apply the axioms and rules of Fig. 1.3 to arrive at the final type for F as follows:

$$F : \{(n, ?), (fac, ?)\} \rightarrow \{(n, 4), (fac, 3)\}$$

which confirms to the results of Fig. 1.2, Sect. 1.1.3.1, which shows that $Rin(1) = (n, ?), (fac, ?)$ and $Rout(4) = (n, 4), (fac, 3)$.

Alternatively, a *type inference algorithm* may be constructed to infer the types of language entities from the types of their surrounding context. The set of typing axioms and rules can then be used to make *judgements* about whether a particular program construct is well-typed within the typing environment or not. Unification in the algorithm may also be used to generate principal types. Whenever all the programs of a language are shown to be well-typed (by proving a type soundness theorem stating that well-typed programs are well behaved), the language is then said to be *type sound*.

Using *static type checking*, the layout of types in a program is tested against the typing rules to reveal any violations. Programs that do not enforce the typing rules and that contain *trapped errors* (errors detected by the typing system) are eliminated. Again, there is an element of approximation in the manner that static type-checking works: if a program passes the type-checker, it is surely well-typed. Otherwise, it is ill-typed or cannot guaranteed to be well-typed. Since defining the typing system is a separate problem from constructing a type-checking algorithm, it is often the case that type system designers run into the problem of defining type systems that will only admit infeasible algorithms or sometimes, no algorithms at all.

Most type systems found in procedural languages are *first order*, i.e. they lack type parameterisation and abstraction. Whenever these are present, the system is called *second order*. Type parameterisation refers to those programs that are of the type $\lambda X.M$, where M is the program and X is a type variable. This means that M is parameterised with respect to X, which is instantiated only by the context. Parameterisation is a feature often found in polymorphic languages, like ML. *Type abstraction* is a feature that often appears as opaque types in interfaces in modular languages, like CLU [37].

Type systems can also use special annotations, called *effects*, that express the effect of using each of the types defined by the typing environment. For example, a function type of the form $\tau_1 \xrightarrow{\varphi} \tau_2$ maps values of type τ_1 to values of type τ_2 and as a result of this application, it may perform a call to function φ. Effect systems are often implemented as extensions of type inference algorithms and are useful in providing information about the internal steps of computation for each program expression.

Finally, one of the interesting topics in the subject of type systems is that of *type equivalence*. Type equivalence relates separately written type expressions. *By-name* equivalence refers to those types that match by having the same names. On the other hand, *structural* equivalence refers to those types that match by having the same structure. It is usual to find typing systems that adopt a mixture of both equivalence relations.

1.2 Literature

We give here an overview of some of the current state of the art relevant to the scope of this book. We shall concentrate on the following main areas: static analysis techniques with applications in the area of security of mobile and cryptographic systems, denotational semantic models for nominal calculi and works related to the formal modelling and analysis as applied to each of the three case studies covered in the following chapters.

1.2.1 Static Analysis Techniques for Program Security

In recent years, the use of static program analysis in analysing mobile and crypto-graphic systems and verifying their security properties has grown into a major area of research, where a variety of techniques have been proposed and implemented to varying degrees of success. In this section, we review research carried out in the areas of control flow analysis, abstract interpretation, type systems and other logic-based areas.

1.2.1.1 Control Flow Analysis

This approach has been adopted on a number of occasions by [38–41], in which the flow logics approach [42] is used to predict certain security properties pertaining to mobile systems, like having the confinement of private information and the adherence to the Bell and LaPadula properties [43].

The main analysis focuses on the usage of channels and the values sent over them in the π-calculus. In particular, a fresh name is associated with a superset of the set of names that can be communicated over that name. Also, an input parameter is associated with a superset of the set of names that may substitute it at runtime.

One major drawback with the manner in which the analysis is carried out is that it identifies all the copies of a fresh name arising from the restriction operator. This implies that non-uniform properties of names that change between the different runs of systems cannot be expressed straightforwardly. The analysis also suffers from the inability to detect certain deadlocks arising from situations where the channel of communication is a fresh name with restricted scope.

The separation between the problem of validating a given solution and the con-struction of such a solution as formulated by the flow logics approach meant that the resulting analyses of [38, 39] are targeted towards the validation of a proposed solution rather than offering a constructive algorithm. However, a minimum solution for the validating analysis is proven to always exist, and in [40], a constructive algorithm for this minimum solution is also supplied assuming a finite set of names.

The security properties dealt with in [38, 39] are all based on the main control flow analysis. In [38], the confinement problem of private names is the main property. In [39], the no read-up/no write-down property introduced by Bell and LaPadula [43] is investigated. This property is often regarded in the security literature as a rigid form of security.

Finally, [41] provides a control flow analysis for a variant of the spi calculus with history-dependent cryptography using the idea of confounders in encrypted messages. This facilitates the comparison of ciphertexts since, for example, $\{0, (\nu r)\}_k \neq \{1, (\nu r)\}_k$. The paper establishes two main results. First, it shows that the Dolev–Yao definition of secrecy [44] is expressible in terms of a control flow analysis. No specification of the Dolev–Yao most general attacker is provided; instead, it is computed using an approximation relation. The second result establishes that the analysis can capture the form of non-interference stated in [45]. This result elegantly separates between the issues of confidentiality and non-interference using the control flow analysis approach.

1.2.1.2   Abstract Interpretation

This approach has been applied effectively to the analysis of mobile systems. Notable works include [46–51]. In [46], a sound non-uniform description of how topologies evolve in closed systems that do not contain nested replications is presented. The approach is based on the abstract data structures of [52], where the abstract meaning of a process in the π-calculus is represented as an undirected *hypergraph* (a hypergraph is a graph where an edge can connect more than two vertices) signifying the sequence of internal computations that evolve within a process from its initial specification. The analysis was further developed and greatly simplified in [47], where a refined semantics is introduced to capture the instance of a channel that establishes a link between two processes.

The study of system interactions in [47] has inspired the analysis of [49], where the main theme is detecting the leakage of confidential information in the presence of unknown contexts. The analysis is non-uniform and can distinguish between the different instances of a process. The open semantics takes into account the lack of specification of the intruder processes and unlike [46, 47], the analysis is not restricted to systems with no nested replications. Encoding the ability to deal with unknown specifications into the semantics is a different alternative to our approach, which adopts a specification of the intruder (i.e. the Dolev–Yao most general attacker).

The analysis of [49] was extended in [50] in an occurrence counting analysis for detecting the exhaustion of resources, mutual exclusion and deadlock properties. On the other hand, a more generic parametric framework has been proposed in [51], which is capable of expressing the equality and inequality relations between names, i.e. the dependency among names at their creation point. The abstract interpretation approach has been used by [48] to approximate cryptographic protocols specified in a cryptographic language with finite principals and sessions (the spi calculus allows infinite principals and sessions) using tree automata. An implementation of the analysis exists. However, it has been applied to small protocols with finite runs only (mostly single runs).

1.2.1.3   Type Systems

Type systems have been extensively researched within the mobile systems and security community and notable examples include the works of [45, 53–58], where properties related to information privacy, resource access control and trust have been tackled.

In [53], a typing system is suggested for a distributed variant of the π-calculus called Dπ, where processes reside at named locations and communications occur locally. In order to be able to express remote communications, a process movement primitive is introduced. The typing system introduces the notion of *location types*, where each location is typed by the resources available to agents residing at that location. This will control the set of capabilities that an agent can perform at a certain location. For example, the agent may be able to send but not receive data. Runtime errors are then used to express the breach of type safety. These errors can detect illegal movements of processes as well as illegal communications.

In [54], dynamic type checking is used to deal with malicious unspecified agents moving between different sites. Open secure semantics are given for the language of Dπ, where special *bad* types are used to distinguish untyped code in what is known as a partial typing system, where only a subset of agents is well-typed. Filters are used by sites to judge the trustworthiness of an incoming agent. Based on this judgement, a notion of authority from which trust between the different sites harbouring processes can be built. Hence, the need for dynamic type checking is, in fact, reduced. The correctness of the behaviour of well-typed processes is shown to be safe with respect to well-behaving sites.

A more extensive treatment of the subject of trust in mobile systems modelled by the π-calculus is given in [58]. Here, a system of Boolean annotations is used to guarantee that only trusted data are used in trusted contexts. The system relies on a notion of trust during runtime. An algorithm for creating the general types of the system is also suggested.

In [56], two security properties, the resource access control and information flow properties, are dealt with in a typing system for a variant of the asynchronous π-calculus [59, 60] with security levels. Using a security policy that assigns levels to processes, a process can only access resources that were created by processes running at the same or lower levels. Similarly, a process can only write to resources created by higher level processes. On the other hand, the implicit flow of information is dealt with by a notion of non-interference, which depends on a form of may-testing.

An advanced static type-checking system is used in [57] to guarantee secure information flow for general process behaviour in the polyadic π-calculus extended with extra syntactic constructs. In this typing system, non-linear types are used as part of a finite partially ordered set of *action types*. The action type of a process is an abstraction of the causal dependency among the free channels of that process. With the use of secrecy indices, a subject reduction property of the resulting typed processes is given. This implies that composing a typed process with external processes does not change its internal secrecy. The subject reduction property is then used to express a form of non-interference among low-level and high-level processes. The resulting system is also used to embed the system of [61], using typed process representation.

One of the major drawbacks of type systems is their inability to express non-uniform properties due to the fact that they associate the same type with the different runs of the system. In [55], this problem was remedied by introducing a primitive for fresh type creation. Special channel types called *Groups* can be created, where names of a type belonging to a particular group cannot be communicated over channels of another group type (possibly with a lower secrecy level). This will protect against the leakage of secret channels from one group to another.

It seems, to date, that the work of [45] has assumed an authoritative role in the subject of security and types in the spi calculus. This is simply due to the wealth of basic concepts that the work contains, and in particular, the idea of mixing (primitively) typed and untyped data in the typing system and the idea that every participant in a security protocol must be typable. An overall view of secrecy principles is also given in the context of the spi calculus. The proposed typing system guarantees the protection of secret data against leakage through a notion of non-interference that is based on testing equivalence. However, the system does not clarify the issue of principal types and therefore, no complete and sound typing algorithm is provided.

Non-interfering security has also been treated in the area of type systems, although to a lesser extent. The work of Smith and Volpano [61] presents a type system for a sequential imperative language that can deal with explicit, as well as implicit, information flow. Insecure explicit flow is detected as a flow from high-level to low-level variables. Implicit information flow, on the other hand, occurring within conditional and while-loop statements is formalised by a type soundness requirement that guarantees that the command levels of the branches of conditional statements and while-loop bodies are as high as the level of the guards of those statements. However, the type system fails in situations where programs do not terminate successfully as a result of assuming that the sequential composition of typable commands is also typable. This problem was later remedied in [62], and a type system for a multi-threaded imperative language was suggested. However, the resulting solution is, unfortunately, too restrictive; the guard of a while-loop is always limited to a low level of security and the body of the while-loop itself is classified as a low-level command. In general, both works [61, 62] adopt a notion of non-interference, which is closely related to the original view given in [63]: the final values of the low-level variables should be independent of the final values of the high-level variables.

A more refined type system for non-interference within concurrent languages is proposed in [59, 64]. The main work extends the system proposed in [62]. However, the attitude towards the concurrent language semantics is that of a small step as compared to the big step semantics adopted by [62]. The implication of this is that intermediate values of low-level variables are also considered (not only the final values). This further results in a stronger notion of non-interference that relieves restriction on the levels of guards and bodies of conditional and while-loop statements. The main difference between [59] and [64] lies in dealing with scheduling mechanisms of concurrent subprocesses (threads).

A simple treatment of non-interference is also given by [65] for an extension of the π-calculus, which can express the independent execution of parallel processes. The definition of non-interference adopted in [65] is based on weak barbed reduction congruence [3]. The semantics of the language annotates processes with security levels drawn from a complete lattice. The final non-interference result then states that pruning high-level sub-terms does not affect the low-level sub-terms up to the weak barbed congruence.

Another extension of the asynchronous π-calculus, termed the security π-calculus, is suggested by [66] that incorporates a type system where security levels and read/write capabilities are associated with input parameters and restricted names. May- and must-testing equivalences are then formalised for the language and based on these equivalence formalisations, notions of non-interference are enforced. The work is a comprehensive extension of the earlier work presented in [56].

Finally, the *compositional security checker* (CoSeC) developed in [67] is a static analysis tool that builds on the *Concurrency Workbench* [68]. Both tools are forms of semantics-based model checkers that can check for properties of processes based on different equivalences and preordering. The language of input for CoSeC is an extension of CCS with separately marked high-level and low-level observable actions. The treatment of the tool benefits only finite state systems and is targeted towards the checking of the non-deducibility-on-compositions property.

1.2.1.4   Other Approaches

Other approaches that cannot be directly classified under the previous headings or that may cover more than one approach have also been suggested and implemented. The work of [69] is an early attempt to present an abstract binding-time analysis as part of a meta-interpreter for the π-calculus. The analysis adopts a partial evaluation technique, which aims at using the known part of the input data in specialising programs. The results of the analysis can determine the static communications that can be executed at compile-time as well as the set of static variables that do not change during runtime. The main motivation behind the work of [69] was program optimisation rather than security (although the results are general enough to include some notion of security).

Mobility Workbench [70] is an example of automated tools that are targeted towards the checking of process equivalences. In particular, the tool checks for the open bisimulation property in the π-calculus [3].

In [71], a typing system is given for a generic version of the spi calculus with constructor/destructor functions. Untyped predicate logic is also used to program a Prolog-based protocol checker that verifies secrecy properties. The methodology is extended, in [72], to check for the authenticity property. The equivalence of the two approaches is established. In [73, 74], a trace analysis is built over a symbolic version of the operational semantics of the spi calculus, and in [75], a well-crafted approach to the symbolic analysis of cryptographic processes, in general, is formalised. Earlier attempts to use symbolic semantics in the analysis of security protocols include [76, 77], where the reachability problem in the cryptographic protocols is discussed.

The works of [78] and [79] have concentrated on the use of automatic checking of equivalences of processes in the spi calculus. The former presents a framework for a trace-based analysis of cryptographic protocols, whereas the latter defines a new equivalence, known as fenced bisimilarity, which removes some of the infinite quantifiers associated with testing equivalence in the spi calculus.

Finally, we mention the works of [80] and [81]. In [80], belief logics have been used to analyse a number of cryptographic protocols and the paper is a notable piece of work. The model-checking approach has also been used with the Brutus checker for a dialect of the spi calculus in [81]. More recently, the Mobility Model Checker (MMC) has been introduced in [82], as a model checker for the π-calculus, which uses tabled logic programming.

1.2.2 Denotational Semantics of Nominal Calculi

Apart from the original structural operational semantics that was proposed for the π-calculus [2], denotational semantic models have also been introduced over the past few years. Some of these include the works of [83–89]. The majority of the work on denotational semantics of mobile systems has adopted an approach based on category theory. The obvious reason underlying this choice seems to be related to the fact that category theory offers abstract reasoning that can be easily utilised in deriving fully abstract denotational models.

The models of [86] and [89] are the first attempts to define a denotational meaning for processes in the π-calculus. Both are fully abstract with respect to late bisimilarity and equivalence. The major difference between the two is that [86] defines the semantics in terms of an internal meta-language for the main category. The language is, in fact, a type-equational theory based on a typed λ-calculus with non-deterministic summation and recursion. The interpretation of processes is then given by translating into this meta-language. This renders the model of [86] a general framework that relies on a uniform approach suitable for modelling a variety of nominal calculi. For example, it is possible to add an interrupt operator to the π-calculus. On the other hand, the model of [89] is more abstract and concentrates on the properties of the categorical solution. This is quite interesting when dealing with the static analysis framework, as it offers concrete grounds on which the analysis can be built. The semantic domains were concretely defined in [89].

Stark's model was further refined in [85] using presheaf categories. These are more flexible as they support function spaces and offer models for synchronisation trees and event structures. Unlike the models of [86, 89], the resulting model is capable of expressing early, as well as late, semantics and hence, it is fully abstract with respect to early bisimilarity and equivalence. Since our static analysis framework is not designed to necessarily deal with the early version of the semantics, we opted for the less elaborate model given in [89].

Two set-theoretic denotational models are presented in [88] that are fully abstract with respect to may- and must-testing in the π-calculus. The models are obtained as solutions of domain equations in a functor category and are quite close to the works of [86, 89]. A higher order version of the π-calculus, where functions as well as names can be communicated over channels, is modelled denotationally in [87]. An internal language is also given for the solution $\mathcal{O}$-category, which is used in the process interpretation. The model is only proven to be fully abstract for transitions (computations) in the higher order language.

In [83], a new notion of Cartesian closed double categories is developed from the theory of Cartesian closed categories and used as a basis for the interpretation of the π-calculus. λ-notation is also used to develop a type system for the resulting interpretation. A final semantics in the style of [90] is also developed using categorical modelling in [91]. The resulting model gives a higher order presentation of the π-calculus using the Edinburgh Logical Framework (LF) (based on typed λ-calculus) as the meta-language. Strong and weak late semantics are both dealt with. However, the full abstraction of the model is not shown with respect to the corresponding bisimilarity and congruence relations. The result is only shown for transitions.

Finally, apart from the category-based models, in the work of [84], a Petri net semantics is developed for the π-calculus with recursive definitions [3]. The semantics is based on place/transition nets with inhibitor arcs and is proven to be sound with respect to early transition semantics of the π-calculus. The model also gives a definition of non-interleaving semantics, which are used to generate causal trees.

1.2.3 Formal Modelling and Analysis of Modern Protocols

We now give an overview of some works in the literature most relevant to the example protocols analysed later in our three case studies.

1.2.3.1 Modelling and Analysis of IoT Protocols

Publish/subscribe is increasingly becoming an important communication paradigm [92], in particular within the domain of sensor device networks and the IoT where messages can be communicated with more efficiency and less consumption of the devices' limited computational power. IBM's MQTT-S protocol [93] was one of the first industrially backed lightweight publish/subscribe protocols that was deployed for wireless sensor and actuator networks. This was followed in the year 2010 by version 3.1 [94], which is currently undergoing standardisation by the OASIS community.

There has been little effort in applying formal analysis tools to IoT communication protocols, mainly due to the novelty of such protocols and their very recent arrival at the scene of communication protocols. On the other hand, some work has been done in the area of publish–subscribe protocols in general. An early attempt in [95] was made to model formally publish/subscribe protocols to capture their essential properties such as minimality and completeness, however, without any attempt to incorporate hostile environments within which these protocols may run.

In [96], the authors define a formal model of publish/subscribe protocols for Grid computing based on Petri nets. Their model offers a mechanism for the composition of existing publish/subscribe protocols with a model, hence offering a friendly approach for the validation of such protocols. Nonetheless, the focus of their work is mostly on Grid computing scenarios. The work of [97] is an early attempt in discussing security properties and requirements desirable in publish/subscribe protocols, in particular within the domain of Internet-based peer-to-peer systems, where such protocols became popular in their early forms.

Several works, e.g. in [98–102], have adopted model checking as an automated technique for verifying properties of systems to verify properties related to reliability and correctness within the context of publish–subscribe systems, and various levels of efficiency. [98, 101] define a general framework for model-checking publish–subscribe systems, without focusing on specific systems or properties. Their approach is more general than ours, though our approach is more efficient since it handles only one class of pub–sub systems (MQTT) and targets one type of properties. The approach of [99] is closer to our approach in that they adopt a specific system, *thinkteam*, as the target for their analysis. In [100], the authors propose a dedicated model-checking technique to verify properties of publish/subscribe-based Message Oriented Middleware (MOM) systems.

Another close work similar to ours is that of [102], where probabilistic model checking is used to capture uncertainties inherent in publish–subscribe systems. However, despite also being stochastic in nature and hence similar to the formal language we adopt here, a major difference is that they use probabilities rather than time as a means of expressing that some communications *probably may not* take place. On the same note, probabilistic model checking is also used in [103] to analyse the quality of predictions in service-oriented architectures.

Within the domain of sensor network protocols, there is more focus of effort on the formal analysis and verification of such protocols. For example, in [104], the authors apply model-checking techniques in the verification of a medium access control protocol called LMAC. Similarly, in [105], the authors propose a formal model of flooding and gossiping protocols for analysing their performance probabilistic properties. More recently, [106] proposed a formal model and analysis of clock-synchronised protocols in sensor networks based on timed automata.

1.2.3.2   Modelling and Analysis of Industry 4.0 Protocols

Formal specification and analysis techniques have been used extensively ever since their inception as a method for increasing the dependability of systems and deepening our understanding of the behaviour of critical systems. Industry 4.0, which brings together several paradigms including Cyber-Physical Systems, Internet of Things (IoT), Cloud Computing, Big Data and Artificial Intelligence among others, poses particular challenges due to the high dimensionality and complexity of the systems involved. Formal methods have much to contribute to the demystification of such complexity [107]. In fact, the important role that formal methods can play in industry to enhance dependability and reliability of industrial systems was emphasised in works as early as [108–110], needless to say that this strong relationship emerged between the two worlds in the form of infamous methods like VDL [111], Z [112] and B [113]. Examples of early surveys highlighting cases where formal methods contributed to the enhancement of industrial systems include [114–118].

More recently, IoT protocols have had a direct relationship with industrial systems, both together sometimes referred to by the term Industrial IoT (IIoT). Among the most widely modelled and formally verified IoT protocols is the MQTT protocol [119]. Various works, e.g. [120–124], have addressed issues with the reliability and security of the protocol using a variety of formal methods. Most notably, in [120, 121], it was discovered that the informal semantics of the protocol contained subtle ambiguities that undermined the protocol's correctness. Other IoT protocols that have benefited from the application of formal methods include CoAP [125], which has been verified in a number of works including [126], who used the Event Calculus [127] to generate a set of monitoring events for the runtime verification of CoAP systems. By contrast, [128] used model-checking techniques to analyse the flow of messages among CoAP nodes.

Apart from this work, other works have also addressed the problem of formally modelling the properties of Industry 4.0 communication technologies. In [129], the authors studied some of the security properties of the Open Platform Communications (OPC) Unified Architecture (UA) [130] using the ProVerif static analysis tool [131]. One of the weaknesses of such tools, in addition to the complexity underlying their usage, is their reliance on the presence of explicit cryptography in the protocol in order to formalise security properties. As such, [132] used UML to semi-formally model UA.

In [133], the authors also used UML, particularly UML activities [134], as part of the Reactive Blocks [135, 136] model-driven approach to model control software in industrial automation. Similarly, UML was combined with net conditions/event systems [137] to model and validate automation systems in [138], particularly plants' structures and dynamics. The formal specification part facilitated the checking of properties such as the lack of dangerous situations in the plant, robustness of the system in terms of the malfunctioning of sensors, avoidance of deadlocks and particularly of relevance to our work in this book, the presence of ample checkpoints in any possible scenario of behaviour the system may follow.

Security has also been the subject of the application of formal verification techniques in the context of Industry 4.0 systems, particularly when it comes to the issue of the integration of various system components. In [139], for example, the Shibboleth federated identity and single sign-on protocol [140] was proposed as a solution to securely connect Fog clients and Fog nodes [141] in automation systems. High-level Petri nets [142] were then used to demonstrate the reliability and robustness of this connection. The various security issues that can arise in Industry 4.0 systems were highlighted in [143].

The approach adopted in this case study follows from a number of existing works [120, 120, 121, 144], which demonstrated the effectiveness of the method in analysing properties and detecting issues with different systems and protocols. The original theory underlying the abstract analysis framework was defined in [145], to which we refer the reader for further detail.

1.2.3.3 Modelling and Analysis of Electric Vehicle Charging Protocols

Despite the importance of modern-day smart and renewable energy systems, their research as an area of application for formal modelling and verification has remained somehow limited, perhaps due to the complexity of such systems. There are a few exceptions. Patil et al. [146] proposed a model-checking framework based on SMV [147] for the purpose of designing robust smart grid applications that check for systems' liveliness and perform boundary value checks. Abate [148], on the other hand, discusses the problem of the verification of networks of smart energy systems and suggests several challenges including the transition from verification of these systems to correct-by-design synthesis and the merger of data-driven approaches with classical formal verification. His paper perhaps best highlights the complexity of such systems, as the main challenge.

Another interesting area that has been used as a case study for formal methods is micro-grids, which represent decentralised areas of electricity generation and consumption. In [149], Gentile et al. propose fluid stochastic Petri nets [150] as a method for modelling the configuration of localised user energy consumption and costs, and present a tool called μGRIMOIRE that can be used in verifying such configurations. In [151], the authors use linear temporal logic as the underlying formalism to specify the requirements of frequency regulation in power systems in relation to the integration of electric vehicles to these systems. The symbolic controller approaches adopted in [151] guarantee the settlement of an after-event's frequency in some specified safe interval, once a failure event is encountered.

Of specific relevance to our case study are works that have targeted the formal modelling and verification of electric vehicle charging protocols, and there have been varieties of such approaches. In particular, we mention [152], in which the authors use the applied π calculus [153] and the ProVerif tool [131] to model and verify privacy properties in a privacy-preserving variant of the vehicle-to-grid communication standard, ISO15118 [154], called POPCORN [155]. Their work, which advocates a Dolev–Yao model of the attacker [44], reveals that while POPCORN preserves weak secrecy, it does not preserve strong secrecy and strong unlinkability, e.g. where a user is linked to their use of the system. The authors also suggest some modifications to the protocol to address these issues.

In other relevant works, dos Santos et al. [156] highlighted the importance of formal modelling in understanding the nature and impact of external threats to vehicular communications and automotive systems. Their work focuses on modelling attacks with the use of predicate/transition nets [157], in particular, attacks related to the compromising of systems' integrity when injected with incorrect data, and whether redundancy countermeasures can help mitigate against such attacks. In [158], the authors combined both the ProVerif [131] and the Tamarin [159] tools to conduct a symbolic analysis of an electric vehicle charging protocol proposed by [160]. Unfortunately, the latter work provides an example of where the application of formal analysis techniques does not rise beyond being a simple exercise with little impact due to the limited usage of the protocols being analysed, even when those protocols are part of an important domain, such as electric vehicle communications. In our case study, we shall seek from the onset to avoid this problem by focusing on (a subset of) a protocol that is widely used at an industrial level. In all of the works above that use formal analysis methods, no mutation testing or mutation generation techniques have been considered. At best, these analyses are only conducted under classical conditions and sometimes assuming the presence of a hostile adversary.

1.3 Further Reading

As we mentioned throughout this introductory chapter, the main areas underlying the background to the research presented in this book can be classified into three main headings: Mobile systems, denotational semantics and static program analysis.

Recommended readings in the area of formalisms for mobile computing include [1], which is an introductory book on the subject of the π-calculus, and [3], which provides a comprehensive treatment of the theory underlying the π-calculus. The main book on the subject of static program analysis is [28], which provides a chapter for each of the main approaches: data flow analysis, control flow analysis, abstract interpretation and type systems. A recommended reading on the subject of denotational semantics is [20], which discusses the denotational methodology in the definition of programming language semantics as introduced by Scott and Strachey [16–18]. A gentler introduction on denotational semantics is given in [21].

References

1. R. Milner, *Communicating and Mobile Systems: The Pi-Calculus* (Cambridge University Press, Cambridge, 1999)
2. R. Milner, J. Parrow, D. Walker, Inf. Comput. **100**(1), 1 (1992)

3. D. Sangiorgi, D. Walker, *The Pi-Calculus - A Theory of Mobile Processes* (Cambridge University Press, Cambridge, 2001)
4. L. Cardelli, A. Gordon, in *Proceedings of the* 1st *International Conference on the Foundations of Software Science and Computation Structures*. Lecture Notes in Computer Science, vol. 1378 (Springer Verlag, Lisbon, Portugal, 1998), pp. 140–155
5. C.A. Petri, Kommunikation mit automaten. Technical report 2, Schriften des Institutes für Instrumentelle Mathematik (1962)
6. C. Hoare, Commun. ACM **21**(8), 666 (1978)
7. R. Milner, Lect. Notes Comput. Sci. **92** (1980)
8. U. Engberg, M. Nielsen, A calculus of communicating systems with name-passing. Technical report DAIMI PB-208, Computer Science Department, University of Aarhus, Aarhus, Denmark (1986)
9. F. Achermann, M. Lumpe, J.G. Schneider, O. Nierstrasz, *Formal Methods for Distributed Processing - A Survey of Object-Oriented Approaches*, pp. 403–426 (2001)
10. C. Fournet, G. Gonthier, in *Proceedings of the 23rd ACM Symposium on Principles of Programming Languages* (ACM Press, St. Petersburg Beach, Florida, USA, 1996), pp. 372–385
11. B. Pierce, D. Turner, Pict: a programming language based on the pi-calculus. Technical report CSCI 476, Computer Science Department, Indiana University (1997)
12. P. Wojciechowski, P. Sewell, in *Proceedings of the* 1st *International Symposium on Agent Systems and Applications, and the* 3rd *International Symposium on Mobile Agents*, eds. by H.E. Bal, B. Belkhouche, L. Cardelli (IEEE Computer Society Press, Palm Springs, California, USA, 1999), pp. 821–826
13. M. Abadi, A. Gordon, in *Proceedings of the* 4th *ACM Conference on Computer and Communications Security* (ACM Press, Zurich, Switzerland, 1997), pp. 36–47
14. J. Vitek, G. Castagna, in *Proceedings of the ICCL Workshop: Internet Programming Languages*, eds. by H.E. Bal, B. Belkhouche, L. Cardelli. Lecture Notes in Computer Science, vol. 1686 (Springer Verlag, Chicago, Illinois, USA, 1998), pp. 47–77
15. R. Milner, The polyadic π-calculus: a tutorial. Technical report ECS-LFCS-91-180, Laboratory for Foundations of Computer Science, Department of Computer Science, University of Edinburgh, UK (1991)
16. D.S. Scott, C. Strachey, in *Proceedings of Symposium on Computers and Automata*, ed. by J. Fox (ACM Press, Polytechnic Institute of Brooklyn, New York, USA, 1971), pp. 19–46
17. D.S. Scott, in *Proceedings of the* 4th *Annual Princeton Conference on Information Science and Systems* (Princeton University, Princeton, New Jersey, USA, 1970), pp. 169–176
18. C. Strachey, *Formal Language Description Languages for Computer Programming*, pp. 198–220 (1966)
19. D. Schmidt, *Denotational Semantics: A Methodology for Language Development* (Allyn & Bacon, Boston, 1986)
20. J.E. Stoy, *Denotational Semantics: The Scott-Strachey Approach to Programming Language Theory* (MIT Press, Cambridge, 1977)
21. G. Winskel, *The Formal Semantics of Programming Languages - An Introduction* (MIT Press, Cambridge, 1993)
22. J. McCarthy, in *Computer Programming and Formal Systems*, eds. by P. Braffort, D. Hirschberg (North-Holland, Amsterdam, Holland, 1965), pp. 33–70
23. J.W. Backus, F.L. Bauer, J. Green, C. Katz, J.L. McCarthy, A.J. Perlis, H. Rutishauser, K. Samelson, B. Vauquois, J.H. Wegstein, A. van Wijngaarden, M. Woodger, P. Naur, Commun. ACM **6**(1), 1 (1963)
24. N. Chomsky, IRE Trans. Inf. Theory **IT-2**(3), 113 (1956)
25. R. Backhouse, *Syntax of Programming Languages: Theory and Practice* (Prentice Hall International, Hoboken, 1979)
26. R.D. Tennent, *Semantics of Programming Languages* (Prentice Hall International, Hoboken, 1991)
27. Pisa Notes (On Domain Theory). www.homepages.inf.ed.ac.uk/gdp/publications/. Accessed 26 Sep 2021

28. F. Nielson, H.R. Nielson, C. Hankin, *Principles of Program Analysis* (Springer, Berlin, 1999)
29. G.A. Kildall, in *Proceedings of ACM Symposium on Principles of Programming Languages* (ACM Press, Boston, Massachusetts, USA, 1973), pp. 194–206
30. T.J. Marlowe, B.G. Ryder, Acta Inform. **28**(2), 121 (1990)
31. J.B. Kam, J.D. Ullman, Acta Inform. **7**, 305 (1977)
32. M. Dwyer, Data flow analysis frameworks for concurrent programs. Technical report 95–8, Kansas State University, Kansas, U.S.A. (1995)
33. L.J. Jagadeesan, R. Jagadeesan, in *Proceedings of the 4th International Conference in Algebraic Methodology and Software Technology*, eds. by V.S. Alagar, M. Nivat. Lecture Notes in Computer Science, vol. 936 (Springer, Montreal, Canada, 1995), pp. 277–291
34. O. Shivers, in *Proceedings of the ACM SIGPLAN'88 Conference on Programming Language Design and Implementation*. ACM SIGPLAN Notices, vol. 23(7) (ACM Press, Atlanta, Georgia, USA, 1988), pp. 164–174
35. P. Cousot, R. Cousot, in *Proceedings of the 4th ACM Symposium on Principles of Programming Languages* (ACM Press, Los Angeles, California, U.S.A., 1977), pp. 238–252
36. J.A. Robinson, J. Assoc. Comput. Mach. **12**(1), 23 (1965)
37. B. Liskov, A. Snyder, R.R. Atkinson, C. Schaffert, Commun. ACM **20**(8), 564 (1977)
38. C. Bodei, P. Dagano, F. Nielson, H.R. Nielson, in *Proceedings of the 9th Conference on Concurrency Theory*. Lecture Notes in Computer Science, vol. 1466 (Springer, Nice, France, 1998), pp. 84–98
39. C. Bodei, P. Dagano, F. Nielson, H.R. Nielson, in *Proceedings of the Conference on Foundations of Software Science and Computation Structures*. Lecture Notes in Computer Science, vol. 1578 (Springer, Lisbon, Portugal, 1999), pp. 120–134
40. C. Bodei, P. Dagano, F. Nielson, H.R. Nielson, Inf. Comput. **168**(1), 68 (2001)
41. C. Bodei, P. Dagano, F. Nielson, H.R. Nielson, in *Proceedings of the 6th International Conference in Parallel Computing Technologies*. Lecture Notes in Computer Science, vol. 2127 (Springer, Novosibirsk, Russia, 2001), pp. 27–41
42. F. Nielson, H.R. Nielson, in *Electronic Notes in Theoretical Computer Science*, vol. 10, eds. by A.P. Andrew Gordon, C. Talcott (Elsevier Science Publishers, Amsterdam, 2000)
43. D. Bell, L.L. Padula, Secure computer systems: unified exposition and multics interpretation. Technical report ESD–TR–75–306, Mitre Corporation, Bedford, M.A., U.S.A. (1975)
44. D. Dolev, A. Yao, in *Proceedings of the 22nd Annual Symposium on Foundations of Computer Science* (1981), pp. 350–357
45. M. Abadi, in *Proceedings of the 3rd International Symposium on Theoretical Aspects of Computer Software*, eds. by M. Abadi, T. Ito. Lecture Notes in Computer Science, vol. 1281 (Springer, Sendai, Japan, 1997), pp. 611–638
46. A. Venet, in *Proceedings of the 5th LOMAPS Workshop on Analysis and Verification of Multiple-Agent Languages*, ed. by M. Dam. Lecture Notes in Computer Science, vol. 1192 (Springer, Stockholm, Sweden, 1996), pp. 51–75
47. A. Venet, in *Proceedings of the 5th International Static Analysis Symposium*, ed. by G. Levi. Lecture Notes in Computer Science, vol. 1503 (Springer, Pisa, Italy, 1998), pp. 152–167
48. D. Monniaux, in *Proceedings of the 6th International Static Analysis Symposium*, eds. by A. Cortesi, G. Filé. Lecture Notes in Computer Science, vol. 1694 (Springer, Venice, Italy, 1999), pp. 149–163
49. J. Feret, in *Proceedings of the 7th International Static Analysis Symposium*. Lecture Notes in Computer Science, vol. 1824 (Springer, University of California, Santa Barbara, USA, 2000), pp. 135–154
50. J. Feret, in *Geometry and Topology in Concurrency theory*, eds. by P. Cousot, E. Goubault, J. Gunawardena, M. Herlihy, M. Raussen, V. Sassone. Electronic Notes in Theoretical Computer Science, vol. 39 (Elsevier Science Publishers, PennState, USA, 2001)
51. J. Feret, in *Proceedings of the 11th European Symposium on Programming*. Lecture Notes in Computer Science, vol. 2305 (Springer, Grenoble, France, 2002), pp. 314–330
52. H. Jonkers, *Algorithmic Languages*, pp. 321–343 (1981)

53. M. Hennessy, J. Riely, in *Proceedings of the* 3rd *International Workshop on High-Level Concurrent Languages*. Electronic Notes in Theoretical Computer Science, vol. 16(3) (Elsevier, Nice, France, 1998), pp. 1–15

54. J. Riely, M. Hennessy, in *Proceedings of the* 26th *ACM SIGPLAN-SIGACT Symposium on Principles of Programming Languages* (ACM Press, San Antonio, TX, USA, 1999), pp. 93–104

55. L. Cardelli, G. Ghelli, A. Gordon, in *Proceedings of the* 11th *International Conference on Concurrency Theory*. Lecture Notes in Computer Science, vol. 1877 (Springer, Penn State University, State College, Pennsylvania, USA, 2000), pp. 365–379

56. M. Hennessy, J. Riely, in *Proceedings of the* 27th *International Colloquium on Automata, Languages and Programming*, eds. by U. Montanari, J.D. Rolim, E. Welzl. Lecture Notes in Computer Science, vol. 1853 (Springer, Geneva, Switzerland, 2000), pp. 415–427

57. K. Honda, V.T. Vasconcelos, N. Yoshida, in *Proceedings of the* 9th *European Symposium on Programming*, ed. by G. Smolka. Lecture Notes in Computer Science, vol. 1782 (Springer, Berlin, Germany, 2000), pp. 180–199

58. M. Hepburn, D. Wright, in *Proceedings of the* 3rd *Conference on Principles and Practice of Declarative Programming* (ACM Press, Florence, Italy, 2001), pp. 103–114

59. G. Boudol, Asynchrony and the π-calculus. Technical report 1702, INRIA-Sophia Antipolis, Sophia Antipolis, France (1992)

60. K. Honda, M. Tokoro, in *Proceedings of the ECOOP'91 Workshop on Object-Based Concurrent Computing*, eds. by M. Tokoro, O. Nierstrasz, P. Wegner. Lecture Notes in Computer Science, vol. 612 (Springer Verlag, Geneva, Switzerland, 1992), pp. 21–51

61. D. Volpano, C. Irvine, G. Smith, J. Comput. Secur. **4**(2,3), 167 (1996)

62. G. Smith, D.M. Volpano, in *Proceedings of the* 25th *ACM SIGPLAN-SIGACT Symposium on Principles of Programming Languages* (ACM Press, San Diego, CA, USA, 1998), pp. 355–364

63. J.A. Goguen, J. Meseguer, in *Proceedings of the IEEE Symposium on Security and Privacy* (IEEE Computer Society Press, Oakland, CA, USA, 1982), pp. 11–20

64. G. Boudol, I. Castellani, Theoret. Comput. Sci. **281**(1), 109 (2002)

65. F. Pottier, in *Proceedings of the* 15th *IEEE Computer Security Foundations Workshop* (IEEE Press, Cape Breton, Nova Scotia, Canada, 2002), pp. 320–330

66. M. Hennessy, The security pi calculus and non-interference. Technical report 05/2000, University of Sussex, Sussex, UK (2000)

67. R. Focardi, R. Gorrieri, IEEE Trans. Software Eng. **23**(9), 550 (1997)

68. R. Cleaveland, J. Parrow, B. Steffen, ACM Trans. Program. Lang. Syst. **15**(1), 36 (1993)

69. M. Gengler, M. Martel, in *Proceedings of the ACM SIGPLAN Symposium on Partial Evaluation and Semantics-Based Program Manipulation* (ACM Press, Amsterdam, The Netherlands, 1997), pp. 36–46

70. B. Victor, F. Moller, in *Proceedings of the* 6th *International Conference in Computer-Aided Verification*, ed. by D. Dill. Lecture Notes in Computer Science, vol. 818 (Springer, Stanford, California, USA, 1994), pp. 428–440

71. M. Abadi, B. Blanchet, in *Proceedings of the* 29th *Annual ACM SIGPLAN-SIGACT Symposium on Principles of Programming Languages* (ACM Press, Portland, USA, 2002), pp. 33–44

72. B. Blanchet, in *Proceedings of the* 9th *International Symposium in Static Analysis*, eds. by M.V. Hermenegildo, G. Puebla. Lecture Notes in Computer Science, vol. 2477 (Springer, Madrid, Spain, 2002), pp. 342–359

73. M. Boreale, in *Proceedings of the* 28th *International Colloquium on Automata, Languages and Programming*, eds. by F. Orejas, P.G. Spirakis, J. van Leeuwen. Lecture Notes in Computer Science, vol. 2076 (Springer, Crete, Greece, 2001), pp. 667–681

74. M. Boreale, M.G. Buscemi, in *Proceedings of the ACM Symposium on Applied Computing* (ACM Press, Madrid, Spain, 2002), pp. 281–285

75. R.M. Amadio, D. Lugiez, V. Vanackére, Theoret. Comput. Sci. **290**(1), 695 (2003)

76. A. Huima, in *Proceedings of the FLOC 1999 Formal Methods and Security Protocols Workshop* (Trento, Italy, 1999), pp. 21–51
77. R. Amadio, D. Lugiez, in *Proceedings of the* 11th *International Conference on Concurrency Theory*, ed. by C. Palamidessi. Lecture Notes in Computer Science, vol. 1877 (Springer, Pennsylvania, USA, 2000), pp. 380–394
78. A. Elkjær, M. Höhle, H. Hüttel, K. Overgård, Combinatorics, Computation & Logic, Australian Computer Science. Communications **21**(3), 175 (1999)
79. V. Cortier, Observational equivalence and trace equivalence in an extension of the spi calculus. Technical report LSV-02-3, Laboratoire Spécification et Vérification, ENS de Cachan, France (2002)
80. M. Burrows, M. Abadi, R. Needham, in *Proceedings of the Royal Society of London A*, vol. 426, pp. 233–271 (1989)
81. S. Gnesi, D. Latella, G. Lenzini, in *Proceedings of the Workshop on Formal Methods and Computer Security* (University of British Columbia, Vancouver, Canada, 2000)
82. P. Yang, C. Ramakrishnan, S. Smolka, in *Proceedings of the* 4th *International Conference in Verification, Model Checking, and Abstract Interpretation*, eds. by L.D. Zuck, P.C. Attie, A. Cortesi, S. Mukhopadhyay. Lecture Notes in Computer Science, vol. 2575 (Springer, New York, USA, 2003), pp. 116–131
83. R. Bruni, U. Montanari, in *Proceedings of the* 14th *Symposium on Logic in Computer Science* (IEEE Computer Society Press, Trento, Italy, 1999), pp. 246–265
84. N. Busi, R. Gorrieri, in *Proceedings of the 6rmth International Conference on Concurrency Theory*. Lecture Notes in Computer Science, vol. 962 (Springer, Philadelphia, PA, USA, 1995), pp. 145–159
85. G.L. Cattani, I. Stark, G. Winskel, in *Proceedings of the* 7th *International Conference on Category Theory and Computer Science*. Lecture Notes in Computer Science, vol. 1290 (Springer Verlag, S. Margherita Ligure, Italy, 1997), pp. 106–126
86. M. Fiore, E. Moggi, D. Sangiorgi, in *Proceedings of the* 11th *Annual IEEE Symposium on Logic in Computer Science* (IEEE Computer Society Press, New Brunswick, New Jersey, USA, 1996), pp. 43–54
87. T. Hartonas, Denotational semantics for a higher-order extension of the monadic π-calculus. Technical report Math&CS1999-1, Technical Education Institute (TEI) of Larissa, Larissa, Greece (1999)
88. M. Hennessy, Theoret. Comput. Sci. **278**(1–2), 53 (2002)
89. I. Stark, in *Proceedings of the* 11th *Annual IEEE Symposium on Logic in Computer Science* (IEEE Computer Society, New Brunswick, New Jersey, USA, 1996), pp. 36–42
90. P. Aczel, in *Proceedings of the* 9th *International Conference in Mathematical Foundations of Programming Semantics*, eds. by S.D. Brookes, M.G. Main, A. Melton, M.W. Mislove, D.A. Schmidt. Lecture Notes in Computer Science, vol. 802 (Springer, New Orleans, USA, 1994), pp. 1–28
91. F. Honsell, M. Lenisa, U. Montanari, M. Pistore, in *Proceedings of the IFIP Working Conference on Programming Concepts and Methods*, eds. by D. Gries, W.P. de Roever. IFIP Conference Proceedings, vol. 125 (Chapman & Hall, Shelter Island, New York, USA, 1998), pp. 21–51
92. A.J. Stanford-Clark, G.R. Wightwick, IBM. J. Res. Dev. **54**(4), 396 (2010)
93. U. Hunkeler, H.L. Truong, A. Stanford-Clark, in *Proceedings of the Third International Conference on Communication System software and Middleware (COMSWARE 2008)* (IEEE, 2008), pp. 791–798
94. D. Locke, MQ Telemetry Transport (MQTT) V3.1 Protocol Specification. Technical report, IBM Corporation (2010)
95. R. Baldoni, M. Contenti, S.T. Piergiovanni, A. Virgillito, in *8th IEEE International Workshop on Object-Oriented Real-Time Dependable Systems (WORDS 2003)* (IEEE Computer Society, 2003), pp. 304–311
96. L. Abidi, C. Cerin, S. Evangelista, in *Proceedings of the 2011 IEEE International Conference on Services Computing* (IEEE Computer Society, Washington, DC, USA, 2011), SCC'11, pp. 496–503

97. C. Wang, A. Carzaniga, D. Evans, A. Wolf, in *Proceedings of the 35th Annual Hawaii International Conference on System Sciences (HICSS'02)-Volume 9 - Volume 9* (IEEE Computer Society, Washington, DC, USA, 2002), HICSS'02, pp. 303–

98. D. Garlan, S. Khersonsky, J.S. Kim, in *Proceedings of the 10th International Conference on Model Checking Software* (Springer, Berlin, Heidelberg, 2003), SPIN'03, pp. 166–180

99. M.H. ter Beek, M. Massink, D. Latella, S. Gnesi, A. Forghieri, M. Sebastianis, Electron. Notes Theor. Comput. Sci. **133**, 275 (2005)

100. Y. Jia, E.L. Bodanese, C.I. Phillips, J. Bigham, R. Tao, in *2014 IEEE Network Operations and Management Symposium, NOMS 2014, Krakow, Poland, May 5-9, 2014* (IEEE, 2014), pp. 1–8

101. L. Baresi, C. Ghezzi, L. Mottola, in *Proceedings of the 29th International Conference on Software Engineering* (IEEE Computer Society, Washington, DC, USA, 2007), ICSE'07, pp. 199–208

102. F. He, L. Baresi, C. Ghezzi, P. Spoletini, in *Formal Techniques for Networked and Distributed Systems - FORTE 2007, 27th IFIP WG 6.1 International Conference, Tallinn, Estonia, June 27-29, 2007, Proceedings*, vol. 4574 (Springer, 2007), pp. 247–262

103. S. Gallotti, C. Ghezzi, R. Mirandola, G. Tamburrelli, in *Proceedings of the 4th International Conference on Quality of Software-Architectures: Models and Architectures* (Springer, Berlin, Heidelberg, 2008), QoSA'08, pp. 119–134

104. A. Fehnker, L.V. Hoesel, A. Mader, in *Proceedings of the 6th International Conference on Integrated Formal Methods* (Springer, Berlin, Heidelberg, 2007), IFM'07, pp. 253–272

105. A. Fehnker, P. Gao, in *Proceedings of the 5th International Conference on Ad-Hoc, Mobile, and Wireless Networks* (Springer, Berlin, Heidelberg, 2006), ADHOC-NOW'06, pp. 128–141

106. F. Heidarian, J. Schmaltz, F.W. Vaandrager, Theor. Comput. Sci. **413**(1), 87 (2012)

107. L.D. Xu, E.L. Xu, L. Li, Int. J. Prod. Res. **56**(8), 2941 (2018)

108. L.M. Barroca, J.A. McDermid, Comput. J. **35**(6), 579 (1992)

109. C.J. Burgess, The role of formal methods in software engineering education and industry. Technical report, University of Bristol, Bristol, UK (1995)

110. P.G. Larsen, J. Fitzgerald, T. Brookes, IEEE Softw. **13**(3), 48 (1996)

111. P. Wegner, ACM Comput. Surv. **4**(1), 5 (1972)

112. J.M. Spivey, *Understanding Z: A Specification Language and Its Formal Semantics* (Cambridge University Press, New York, 1988)

113. J.R. Abrial, *The B-book: Assigning Programs to Meanings* (Cambridge University Press, New York, 1996)

114. D. Craigen, S. Gerhart, T. Ralston, in *Z User Workshop, London 1992* (Springer, 1993), pp. 1–5

115. M.G. Hinchey, J.P. Bowen, *Applications of Formal Methods*, vol. 1 (Prentice Hall, New Jersey, 1995)

116. J. Bowen, V. Stavridou, Softw. Eng. J. **8**(4), 189 (1993)

117. J.R. Abrial, E. Börger, H. Langmaack, et al., *Formal Methods for Industrial Applications: Specifying and Programming The Steam Boiler Control*, vol. 9 (Springer Science & Business Media, 1996)

118. G. Frey, L. Litz, in *SMC 2000 Conference Proceedings. 2000 IEEE International Conference on Systems, Man and Cybernetics. 'Cybernetics Evolving to Systems, Humans, Organizations, and their Complex Interactions'(cat. no. 0*, vol. 4 (IEEE, 2000), pp. 2431–2436

119. A. Banks, R. Gupta, MQTT Version 3.1.1 Plus Errata 01. Technical report, OASIS (2015)

120. B. Aziz, in *9th International Conference on Availability, Reliability and Security (ARES 2014), Fribourg, Switzerland* (IEEE, 2014)

121. B. Aziz, Ad Hoc Netw. **36**, 49 (2016)

122. S. Chouali, A. Boukerche, A. Mostefaoui, in *Proceedings of the 15th ACM International Symposium on Mobility Management and Wireless Access* (ACM, 2017), pp. 129–136

123. K. Mladenov, Formal verification of the implementation of the MQTT protocol in IoT devices. Master's thesis, University of Amsterdam, the Netherlands (2017)

124. M. Houimli, L. Kahloul, S. Benaoun, in *2017 International Conference on Mathematics and Information Technology (ICMIT)* (IEEE, 2017), pp. 214–221
125. C. Bormann, A.P. Castellani, Z. Shelby, IEEE Internet Comput. **16**(2), 62 (2012)
126. K. İnçki, I. Ari, IEEE Access **6**, 13501 (2018)
127. R. Kowalski, M. Sergot, N. Gener. Comput. **4**(1), 67 (1986)
128. A.J. Vattakunnel, N.S. Kumar, G.S. Kumar, Procedia Comput. Sci. **93**, 299 (2016)
129. M. Puys, M.L. Potet, P. Lafourcade, in *International Conference on Computer Safety, Reliability, and Security* (Springer, 2016), pp. 67–75
130. The OPC Unified Architecture. https://opcfoundation.org/about/opc-technologies/opc-ua/. Accessed 11 Jun 2019
131. ProVerif: Cryptographic protocol verifier in the formal model. http://prosecco.gforge.inria.fr/personal/bblanche/proverif/. Accessed 24 Sep 2014
132. S. Rohjans, K. Piech, S. Lehnhoff, in *2013 IEEE International Workshop on Inteligent Energy Systems (IWIES)* (2013), pp. 209–214
133. P. Herrmann, J.O. Blech, in *Federation of International Conferences on Software Technologies: Applications and Foundations* (Springer, 2016), pp. 253–261
134. Object Management Group: OMG Unified Modeling Language (OMG UML), Superstructure. https://www.omg.org/spec/UML/2.4.1/Superstructure/PDF/. Accessed 03 Jun 2019
135. Bitreactive AS: Reactive Blocks. http://www.bitreactive.com. Accessed 03 Jun 2019
136. F.A. Kraemer, V. Slåtten, P. Herrmann, J. Syst. Softw. **82**(12), 2068 (2009)
137. M. Rausch, H.. Hanisch, in *Proceedings 1995 INRIA/IEEE Symposium on Emerging Technologies and Factory Automation. ETFA'95*, vol. 1 (1995), pp. 592–600
138. H.M. Hanisch, A. Lobov, J.L.M. Lastra, R. Tuokko, V. Vyatkin, Int. J. Manuf. Technol. Manage. **8**(1–3), 75 (2006)
139. S. Zahra, M. Alam, Q. Javaid, A. Wahid, N. Javaid, S.U.R. Malik, M. Khurram Khan, IEEE Access **5**, 27132 (2017)
140. Shibboleth. https://www.shibboleth.net/. Accessed 11 Jun 2019
141. F. Bonomi, R. Milito, J. Zhu, S. Addepalli, in *Proceedings of the First Edition of the MCC Workshop on Mobile Cloud Computing* (ACM, 2012), pp. 13–16
142. K. Jensen, in *Applications and Theory of Petri Nets* (Springer, Berlin, 1983), pp. 166–180
143. D. Chen, G. Chang, KSII Trans. Internet Inf. Syst. Comput. Sci.**6**(1) (2012)
144. B. Aziz, in *Proceedings of the 11th International Conference on Algebraic Methodology and Software Technology*. Lecture Notes in Computer Science, vol. 4019 (Springer, Kuressaare, Estonia, 2006), pp. 36–50
145. B. Aziz, A static analysis framework for security properties in mobile and cryptographic systems. Ph.D. thesis, School of Computing, Dublin City University, Dublin, Ireland (2003)
146. S. Patil, G. Zhabelova, V. Vyatkin, B. McMillin, in *IECON 2015 - 41st Annual Conference of the IEEE Industrial Electronics Society* (2015), pp. 003,974–003,979
147. K.L. McMillan, in *Symbolic Model Checking* (Springer, Berlin, 1993), pp. 25–60
148. A. Abate, in *Numerical Software Verification*, eds. by S. Bogomolov, M. Martel, P. Prabhakar (Springer International Publishing, Cham, 2017), pp. 1–14
149. U. Gentile, S. Marrone, N. Mazzocca, R. Nardone, Open Cybern. Syst. J. **10**(1) (2016)
150. M. Gribaudo, M. Sereno, A. Bobbio, in *Proceedings 8th International Workshop on Petri Nets and Performance Models (Cat. No.PR00331)* (1999), pp. 74–81
151. B. Wooding, V. Vahidinasab, S. Soudjani, in *2020 International Conference on Smart Energy Systems and Technologies (SEST)* (2020), pp. 1–6
152. M. Fazouane, H. Kopp, R.W. van der Heijden, D. Le Métayer, F. Kargl, in *Engineering Secure Software and Systems*. ed. by F. Piessens, J. Caballero, N. Bielova (Springer International Publishing, Cham, 2015), pp. 17–33
153. M. Abadi, C. Fournet, in *Proceedings of the 28th ACM SIGPLAN-SIGACT Symposium on Principles of Programming Languages* (Association for Computing Machinery, New York, NY, USA, 2001), POPL '01, p. 104–115
154. I.O. for Standardization. Iso 15118-1:2019 road vehicles — vehicle to grid communication interface — part 1: General information and use-case definition (2019)

155. C. Höfer, J. Petit, R. Schmidt, F. Kargl, in *Proceedings of the 2013 ACM Workshop on Security, Privacy & Dependability for Cyber Vehicles* (Association for Computing Machinery, New York, NY, USA, 2013), CyCAR'13, pp. 37–48

156. E. dos Santos, D. Schoop, A. Simpson, in *2016 IEEE Vehicular Networking Conference (VNC)* (2016), pp. 1–8

157. H.J. Genrich, in *Petri Nets: Central Models and Their Properties*, eds. by W. Brauer, W. Reisig, G. Rozenberg (Springer, Berlin, 1987), pp. 207–247

158. L. Li, J. Pang, Y. Liu, J. Sun, J.S. Dong, in *2014 19th International Conference on Engineering of Complex Computer Systems* (2014), pp. 11–18

159. S. Meier, B. Schmidt, C. Cremers, D. Basin, in *Computer Aided Verification*. ed. by N. Sharygina, H. Veith (Springer, Berlin, 2013), pp. 696–701

160. J.K. Liu, M.H. Au, W. Susilo, J. Zhou, in *Computer Security - ESORICS 2012*, eds. by S. Foresti, M. Yung, F. Martinelli (Springer, Berlin, 2012), pp. 397–414

Chapter 2
Process Algebra: Syntax and Semantics

2.1 Introduction

The concept of a name as the main abstraction underlying different low-level details of computing systems has resulted in the proliferation of a family of models of computation known as *nominal calculi*. Many such models are *process algebraic*; programs are modelled as processes equipped with operations to express their *evolution*. The π-calculus has emerged in recent years as an authoritative and highly expressive nominal calculus for the modelling of mobile systems.

For years, the π-calculus has provided a powerful, yet simple, theoretical framework for modelling and reasoning about mobility. The notion of a *name* constitutes the central concept on which the calculus is based. Names are allowed to stand for channels as well as data. Therefore, by sending and receiving names, processes can communicate information to other processes and can also change the topology of their network. This view of mobility is described as "the movement of links in the virtual space of linked processes" [1, II-8]. Several versions of the π-calculus exist, including the monadic synchronous version, which was originally developed in [2]. In this version, communications occur by synchronising channels and carrying messages that are single names. Other versions exist, notably the polyadic π-calculus [3] and the asynchronous π-calculus [4, 5].

In this chapter, we review the syntax and structural operational semantics of a special variation of the π-calculus with timed input constructs. We give the non-standard meaning of this language based on domain theory and to capture the name substitution property of communications.

2.2 Syntax

The syntax of our process algebra is defined as follows, using the classical Backus–Naur Form (BNF) [6, 7]:

© The Author(s), under exclusive license to Springer Nature Switzerland AG 2022
B. Aziz, *Formal Analysis by Abstract Interpretation*, SpringerBriefs in Applied
Sciences and Technology, https://doi.org/10.1007/978-3-030-91153-9_2

$P, Q ::= \mathbf{0} \mid \overline{x}\langle y \rangle.P \mid \texttt{timer}^t(x(y).P, Q) \mid \tau.P \mid !P \mid (\nu x)P \mid (P \mid Q) \mid (P + Q)$
$\mid A(x) \mid P[x = y]Q$

where processes $P, Q, \ldots \in \mathcal{P}$ form an infinite set. The structure of these is defined
as follows. A null process, $\mathbf{0}$, is an inactive process that cannot evolve any further.
A guarded process may perform an output action, $\overline{x}\langle y \rangle.P$, and then continues as the
residue, P. A guarded process may also be a timed input action, $\texttt{timer}^t(x(y).P, Q)$,
where $t \in \mathbb{N}$ is a natural number representing a time bound on the availability of the
input action. Basically, the input action, $x(y).P$, can synchronise with suitable output
actions, e.g. $\overline{x}\langle z \rangle.P$, as long as $t > 0$, in which case a message z received along x
will substitute y in the residual process, P. This substitution is denoted as $P[z/y]$.
Otherwise, when t reaches 0 and no matching output actions have become available,
the timer will switch to behaving as Q. There is an assumption that t is decremented
by the environment the process is running within, and that t can be any time unit (e.g.
a tick and a second). This can be expressed by means of a clock function, $\eth : \mathcal{P} \to \mathcal{P}$,
that *ticks down* time [8]:

$$\eth(P) = \begin{cases} \texttt{timer}^t(x(y).Q, R) & \text{if } P = \texttt{timer}^{t+1}(x(y).Q, R) \\ & \text{and } 0 < t + 1 < \infty \\ \texttt{timer}^t(x().Q, R) & \text{if } P = \texttt{timer}^{t+1}(x().Q, R) \\ & \text{and } 0 < t + 1 < \infty \\ \eth(Q) \mid \eth(R) & \text{if } P = Q \mid R \\ \eth(Q) + \eth(R) & \text{if } P = Q + R \\ (\nu x)\eth(Q) & \text{if } P = (\nu x)Q \\ !\eth(Q) & \text{if } P = !Q \\ P & \text{otherwise} \end{cases}$$

Sometimes, a communication does not pass any messages (i.e. it is a pure synchronisation à la CCS [9]), in which case we write the output action as $\overline{x}\langle\rangle.P$ and the
corresponding input action as $\texttt{timer}^t(x().P, Q)$.

A silent guard, $\tau.P$, will simply continue as P after some unknown time in which
τ is consumed. This silent guard is often used to model invisible internal activities.
A process can be replicated, $!P$, to generate as many copies of P as required by its
environment. It can also restrict the scope of a name as in $(\nu x)P$, where x will not be
visible outside of P. Two processes, P and Q, can be combined together in parallel,
$(P \mid Q)$, thus interleaving P and Q and allowing for any communications between the
two processes to occur, or can be combined using a non-deterministic choice operator,
$(P + Q)$, where either P or Q is chosen randomly to evolve, dropping the other
process. A process definition call, $A(x)$, calls a process definition, $A(y) \overset{\text{def}}{=} P$, and at
the same time passes the value of x to the process P while replacing y. If the process
definition does not accept any input parameters, we simply write it as $A() \overset{\text{def}}{=} P$.
Finally, the special construct, $P[x = y]Q$, will evaluate the condition, $x = y$, and
if this condition is found to be true, it will continue as process P, otherwise, it will
continue as process Q. We call the set of free names of a process, $fn(P)$, such as x
and y in $\overline{x}\langle y \rangle.P$, $x(u).P$ and $P[x = y]Q$, and the set of bound names, $bn(P)$, such

as u and y in $x(u).P$ and $(vy)P$. The set of names of a process, $n(P)$, is hence equal to the union $fn(P) \cup bn(P)$.

We now proceed to define the semantics of the above language more formally using the classical structural operational semantics approach.

2.3 Structural Operational Semantics

The structural operational semantics of our language is formalised using a *structural congruence* relation, $\equiv$, and a *labelled transition* relation, $\xrightarrow{\mu}$, the rules for both of which are defined in Fig. 2.1.

The structural congruence relation states that a process can change its structure to accommodate any necessary evolution, and therefore, the process is not evolving yet, but only changing its shape. Rule (1) states that a process is a commutative monoid under parallel composition. Rule (2) states that name restrictions can be dropped on a null process. Rule (3) allows for name restriction to be re-ordered. Rule (4) states that name restriction can be passed to one side of parallel composition, as long as the restricted name does not belong to the set of free names of the subprocess. Rule (5) allows for a replicated process to spawn a copy of itself. Rule (6) states that the non-deterministic choice is commutative. Rules (7)–(10) deal with the reshaping of timed inputs, depending on the value of the time bound t, while Rule (11) allows for the clock function to be applied to a timed input process in order to tick the time. Rule (12) calculates the equality of two names, and reshapes the process according to the outcome. Finally, Rules (13) and (14) replace process definition calls by the body of the definition of the process.

The labelled transition relation, on the other hand, expresses an evolution step for a process, in terms of what action (or a better word would be, emitted behaviour), $\mu \in \{\overline{x}\langle y \rangle, \overline{x}\langle \rangle, x(z), x(), \tau\}$, can be observed externally by the environment of that process. Rule (15) expresses the emission of a free output action. Rule (16) expresses the emission of a free output action with no message included. Rules (17) and (18) express live inputs, with and without input messages accepted, respectively. Rule (19) turns a free output emission to a bound output emission, provided that the name restriction does not apply to the name of the channel on which the output is happening. Rules (20) and (21) allow for output-emitting and input-emitting processes to interact and emit a silent action representing an internal communication that may be free (Rule (20)) or bound (Rule (21)). Rule (22) is similar, except the internal communication does not involve any exchange of messages. Rule (23) states that any behaviour emitted by a process can still be emitted under the presence of name restriction, as long as the name being restricted is not in the free names of the emitted action, μ. Rule (24) states that the emitted behaviour is preserved under parallel composition. Rule (25) states that the emitted behaviour allows for the choice operator to evolve choosing the process that performs the emitted action. Finally, Rule (26) allows for a process guarded by a silent action to emit that action and proceed to its residue, P.

Rules of the ≡ relation:

(1) $(\mathcal{P}/\equiv, |, \mathbf{0})$ is a commutative monoid

(2) $(\nu x)\mathbf{0} \equiv \mathbf{0}$

(3) $(\nu x)(\nu y)P \equiv (\nu y)(\nu x)P$

(4) $(\nu x)(P \mid Q) \equiv (P \mid (\nu x)Q)$ *if* $x \notin fn(Q)$

(5) $!P \equiv P \mid !P$

(6) $(P + Q) \equiv (Q + P)$

(7) $\mathtt{timer}^0(x(z).P, Q) \equiv Q$

(8) $\mathtt{timer}^0(x().P, Q) \equiv Q$

(9) $\mathtt{timer}^\infty(x(z).P, Q) \equiv x(z).P$

(10) $\mathtt{timer}^\infty(x().P, Q) \equiv x().P$

(11) $\mathtt{timer}^t(x(z).P, Q) \equiv \eth(\mathtt{timer}^t(x(z).P, Q))$

(12) $P[x = y]Q = \begin{cases} P & \text{if } x = y \\ Q & \text{otherwise} \end{cases}$

(13) $A(x) \equiv P[x/y]$ *where* $A(y) \stackrel{\text{def}}{=} P$

(14) $A() \equiv P$ *where* $A() \stackrel{\text{def}}{=} P$

Rules of the $\xrightarrow{\mu}$ relation:

(15) $\overline{x}\langle y\rangle.P \xrightarrow{\overline{x}\langle y\rangle} P$

(16) $\overline{x}\langle\rangle.P \xrightarrow{\overline{x}\langle\rangle} P$

(17) $\mathtt{timer}^{t+1}(x(z).P, Q) \xrightarrow{x(z)} P$

(18) $\mathtt{timer}^{t+1}(x().P, Q) \xrightarrow{x()} P$

(19) $P \xrightarrow{\overline{x}\langle y\rangle} Q \;\Rightarrow\; (\nu y)P \xrightarrow{\overline{x}\langle y\rangle} Q$ *if* $x \neq y$

(20) $P \xrightarrow{\overline{x}\langle y\rangle} P', Q \xrightarrow{x(z)} Q' \;\Rightarrow\; P \mid Q \xrightarrow{\tau} P' \mid Q'[y/z]$

(21) $P \xrightarrow{\overline{x}(y)} P', Q \xrightarrow{x(z)} Q' \;\Rightarrow\; P \mid Q \xrightarrow{\tau} (\nu y)(P' \mid Q'[y/z])$

(22) $P \xrightarrow{\overline{x}\langle\rangle} P', Q \xrightarrow{x()} Q' \;\Rightarrow\; P \mid Q \xrightarrow{\tau} P' \mid Q'$

(23) $P \xrightarrow{\mu} Q \;\Rightarrow\; (\nu x)P \xrightarrow{\mu} (\nu x)Q$ *if* $x \notin fn(\mu)$

(24) $P \xrightarrow{\mu} P' \;\Rightarrow\; P \mid Q \xrightarrow{\mu} P' \mid Q$

(25) $P \xrightarrow{\mu} P' \;\Rightarrow\; P + Q \xrightarrow{\mu} P'$

(26) $\tau.P \xrightarrow{\tau} P$

Fig. 2.1 The structural operational semantics of the process algebra

2.4   A Non-standard Name-Substitution Semantics

To trace name substitutions during the evolution of processes in our language, we define a non-standard name-substitution semantics that defines a different meaning for a process from the meaning given by the structural operational semantics.

We start first by defining a special environment, $\phi_{\mathcal{E}} : N \to \wp(N)$, which maps each name of a process to a set of names that may instantiate that name during its runtime. The null environment, $\phi_{\mathcal{E}0}$, as such is defined as the environment that maps every name to the empty set:

$$\forall x \in N : \phi_{\mathcal{E}0}(x) = \{\}$$

When computing the meaning of a process, P, elements of the set $bn(P)$, representing input parameters, will obtain values in $\phi_{\mathcal{E}}$ whenever communications take place. For example, if a message, z, is received over a channel, x, in an input action, $\mathtt{timer}^t(x(y).P, Q)$, then z will be added to the set of names held by $\phi_{\mathcal{E}}(y)$. The fact that the semantics is precise, i.e. every instance of a bound name is distinguished from every other instance in the semantics, means that the set $\phi_{\mathcal{E}}(y)$ will be at most a singleton set per the choice of control flow. However, this set is then compounded with the multiple possible substitutions resulting from the different possible choices of control flows.

As a result of these substitutions, a semantic domain $D_\perp = N \to \wp(N)$ is formed with the following partial ordering relation, $\sqsubseteq_{D_\perp}$, based on subset inclusion:

$$\forall \phi_{\mathcal{E}1}, \phi_{\mathcal{E}2} \in D_\perp : \phi_{\mathcal{E}1} \sqsubseteq_{D_\perp} \phi_{\mathcal{E}2} \Leftrightarrow \forall x \in N : \phi_{\mathcal{E}1}(x) \subseteq \phi_{\mathcal{E}2}(x)$$

with the bottom element of $D_\perp$ being the null environment, $\phi_{\mathcal{E}0}$. The union of $\phi_{\mathcal{E}}$ environments, $\cup_{\phi_{\mathcal{E}}}$, can thus be defined in terms of the standard union, $\cup$, as follows:

$$\forall \phi_{\mathcal{E}1}, \phi_{\mathcal{E}2} \in D_\perp, x \in N : (\phi_{\mathcal{E}1} \cup_{\phi_{\mathcal{E}}} \phi_{\mathcal{E}2})(x) = \phi_{\mathcal{E}1}(x) \cup \phi_{\mathcal{E}2}(x)$$

We explain how this non-standard semantics is *precise*. Informally, this means that the semantics preserves copies of names resulting from α-conversions or renaming of bound names. For example, in the process

$$!((\nu y)\overline{x}\langle y\rangle.\mathbf{0}) \mid !\mathtt{timer}^t(x(u).\mathbf{0}, \mathbf{0})$$

our non-standard interpretation now captures precisely the set of substitutions, $\{y_1/u_1, y_2/u_2 \ldots\}$, where y_i is a labelled copy of the fresh name, y, and u_i is a labelled instance of the input parameter, u, assuming that t remains above 0.

From the above definition of $D_\perp$, we can assign a meaning to some process P using the syntax-directed non-standard semantic function, $\mathcal{E}[\![P]\!] \rho \phi_{\mathcal{E}} \in D_\perp$, defined over the structure of processes, as shown in Fig. 2.2.

$(\mathcal{E}1)$ $\quad \mathcal{E}([\overline{x}\langle y\rangle.P]) \, \rho \, \phi_{\mathcal{E}} \qquad\qquad\qquad = \phi_{\mathcal{E}}$

$(\mathcal{E}2)$ $\quad \mathcal{E}([\overline{x}\langle\rangle.P]) \, \rho \, \phi_{\mathcal{E}} \qquad\qquad\qquad = \phi_{\mathcal{E}}$

$(\mathcal{E}3)$ $\quad \mathcal{E}([\texttt{timer}^{t+1}(x(y).P, Q)]) \, \rho \, \phi_{\mathcal{E}} =$

$$(\bigcup\nolimits_{\phi_{\mathcal{E}}} \mathcal{R}([(\biguplus_{R\in\rho} \{|R|\}) \uplus \{|P|\} \uplus \{|P'|\}]) \, \phi_{\mathcal{E}}[y \mapsto \phi_{\mathcal{E}}(y) \cup \{z\}]) \cup_{\phi_{\mathcal{E}}}$$

$$\overline{x'}\langle z\rangle.P'\in\rho:\ x\overset{\phi_{\mathcal{E}}}{\sim}x'$$

$$\mathcal{R}([(\biguplus_{R\in\rho} \{|R|\}) \uplus \{|\texttt{timer}^{t}(x(y).P, Q)|\}]) \, \phi_{\mathcal{E}}$$

$(\mathcal{E}4)$ $\quad \mathcal{E}([\texttt{timer}^{t+1}(x().P, Q)]) \, \rho \, \phi_{\mathcal{E}} =$

$$(\bigcup\nolimits_{\phi_{\mathcal{E}}} \mathcal{R}([(\biguplus_{R\in\rho} \{|R|\}) \uplus \{|P|\} \uplus \{|P'|\}]) \, \phi_{\mathcal{E}}) \cup_{\phi_{\mathcal{E}}}$$

$$\overline{x'}\langle\rangle.P'\in\rho:\ x\overset{\phi_{\mathcal{E}}}{\sim}x'$$

$$\mathcal{R}([(\biguplus_{R\in\rho} \{|R|\}) \uplus \{|\texttt{timer}^{t}(x(y).P, Q)|\}]) \, \phi_{\mathcal{E}}$$

$(\mathcal{E}5)$ $\quad \mathcal{E}([\texttt{timer}^{0}(x(y).P, Q)]) \, \rho \, \phi_{\mathcal{E}} = \mathcal{R}([\{|Q|\} \uplus \rho]) \, \phi_{\mathcal{E}}$

$(\mathcal{E}6)$ $\quad \mathcal{E}([\texttt{timer}^{0}(x().P, Q)]) \, \rho \, \phi_{\mathcal{E}} = \mathcal{R}([\{|Q|\} \uplus \rho]) \, \phi_{\mathcal{E}}$

$(\mathcal{E}7)$ $\quad \mathcal{E}([P \mid Q]) \, \rho \, \phi_{\mathcal{E}} \qquad\quad = \mathcal{R}([\{|P|\} \uplus \{|Q|\} \uplus \rho]) \, \phi_{\mathcal{E}}$

$(\mathcal{E}8)$ $\quad \mathcal{E}([!P]) \, \rho \, \phi_{\mathcal{E}} \qquad\qquad = snd(\textit{fix } \mathcal{F}(0, \bot_{D}))$

where, $\mathcal{F} = \lambda f\, \lambda(j, \phi).$

$$f \ (if \phi = \mathcal{R}([(\biguplus_{i=0}^{j}\{|(P)\sigma|\}) \uplus \rho]) \, \phi_{\mathcal{E}} \ then$$

$$j, \phi \ else \ (j+1), (\mathcal{R}([(\biguplus_{i=0}^{j}\{|(P)\sigma|\}) \uplus \rho]) \, \phi_{\mathcal{E}}))$$

and $\sigma = [bn_{i}(P)/bn(P)]$, $bn_{i}(P) = \{x_{i} \mid x \in bn(P)\}$

$(\mathcal{E}9)$ $\quad \mathcal{E}([(\nu n)P]) \, \rho \, \phi_{\mathcal{E}} \qquad = \mathcal{R}([\{|P|\} \uplus \rho]) \, \phi_{\mathcal{E}}$

$(\mathcal{E}10)$ $\mathcal{E}([0]) \, \rho \, \phi_{\mathcal{E}} \qquad\qquad = \phi_{\mathcal{E}}$

$(\mathcal{E}11)$ $\mathcal{E}([P + Q]) \, \rho \, \phi_{\mathcal{E}} \qquad = \mathcal{E}([P]) \, \rho \, \phi_{\mathcal{E}} \ \cup_{\phi_{\mathcal{E}}} \ \mathcal{E}([Q]) \, \rho \, \phi_{\mathcal{E}}$

$(\mathcal{E}12)$ $\mathcal{E}([A(x)]) \, \rho \, \phi_{\mathcal{E}} \qquad = \mathcal{E}([P[x/y]]) \, \rho \, \phi_{\mathcal{E}} \ where, A(y) \overset{\text{def}}{=} P$

$(\mathcal{E}13)$ $\mathcal{E}([A]) \, \rho \, \phi_{\mathcal{E}} \qquad\qquad = \mathcal{E}([P]) \, \rho \, \phi_{\mathcal{E}} \ where, A() \overset{\text{def}}{=} P$

$(\mathcal{E}14)$ $\mathcal{E}([\tau.P]) \, \rho \, \phi_{\mathcal{E}} \qquad\quad = \mathcal{E}([P]) \, \rho \, \phi_{\mathcal{E}}$

$(\mathcal{E}15)$ $\mathcal{E}([P[x = y]Q]) \, \rho \, \phi_{\mathcal{E}} \quad = \begin{cases} \mathcal{E}([P]) \, \rho \, \phi_{\mathcal{E}} & \textit{if } x \overset{\phi_{\mathcal{E}}}{\sim} y \\ \mathcal{E}([Q]) \, \rho \, \phi_{\mathcal{E}} & \textit{otherwise} \end{cases}$

$(\mathcal{R}0)$ $\mathcal{R}([\rho]) \, \phi_{\mathcal{E}} \qquad\qquad = \bigcup_{\phi_{\mathcal{E}}}_{P\in\rho} \mathcal{E}([P]) \, (\rho \backslash \{|P|\}) \, \phi_{\mathcal{E}}$

Fig. 2.2 The definition of the non-standard semantic function, $\mathcal{E}([P]) \, \rho \, \phi_{\mathcal{E}}$

In the rules of this non-standard semantics, ρ is a multiset (bag) of processes running in parallel with the interpreted process along with the standard operators, $\{| - |\} : \mathcal{P} \to \wp(\mathcal{P})$ and $\uplus : \wp(\mathcal{P}) \times \wp(\mathcal{P}) \to \wp(\mathcal{P})$, for multiset creation and multiset union, respectively, defined for ρ. The meaning of a multiset of processes, ρ, is given in the rule, $(\mathcal{R}0)$, using the special union, $\cup_{\phi_{\mathcal{E}}}$, defined earlier. We describe next the rest of the semantic rules informally. Rule $(\mathcal{E}1)$ for output actions with messages does not affect the value of $\phi_{\mathcal{E}}$ since communications are dealt with in the case of input actions (i.e. in rule $(\mathcal{E}3)$). In rule $(\mathcal{E}2)$, again no communications take place and hence no name substitutions are captured. In rule $(\mathcal{E}3)$ for active input actions, the rule uses the equivalence of a two-name relation, $\overset{\phi_{\mathcal{E}}}{\sim}$, parameterised by $\phi_{\mathcal{E}}$, to determine matching channel names. This relation is defined for any two names, x and y, as follows:

$$x \overset{\phi_{\mathcal{E}}}{\sim} y \quad \Leftrightarrow \quad (\phi_{\mathcal{E}}(x) \cap \phi_{\mathcal{E}}(y) \neq \emptyset) \vee (x = y)$$

For each message-passing synchronisation, the value of $\phi_{\mathcal{E}}$ is updated with the communicated message. The rule also takes into account the choice of no communication taking place, which then leads to simply the ticking of the clock for the timer. The next rule, $(\mathcal{E}4)$, deals with a similar case except no message is being passed by the matching outputs. The difference here is that the value of $\phi_{\mathcal{E}}$ will remain unchanged in the event of synchronisation taking place. Rules $(\mathcal{E}5)$ and $(\mathcal{E}6)$ deal with inactive input actions, where the timer has expired (i.e. reached zero value). The meaning of the input-guarded process becomes the meaning of the alternative continuation process, Q. The two cases cover input expecting a message and those not expecting a message.

Rule $(\mathcal{E}7)$ naturally adds two parallel processes in the ρ multiset. Rule $(\mathcal{E}8)$ deals with replicated processes using a fixed-point calculation of the higher order functional, $\mathcal{F}$. The rule allows as many copies of P to be spawned, and the number of each copy is used to subscript its bound names in order to maintain their uniqueness. As a result, the interpretation of restricted names in rule $(\mathcal{E}9)$ drops the ν operator in ρ, since this becomes unnecessary in our semantics due to name uniqueness. Due to the infinite nature of the non-standard semantic domain, $D_{\perp}$, computing the least fixed point for $\mathcal{F}$ is not guaranteed to terminate. Rule $(\mathcal{E}10)$ returns the same value of $\phi_{\mathcal{E}}$ when interpreting the null process, and in rule $(\mathcal{E}11)$, the union of the two possible choice interpretations is given as the interpretation of the choice operator. Rules $(\mathcal{E}12)$ and $(\mathcal{E}13)$ replace a process application with its definition making the necessary name substitutions (rule $(\mathcal{E}12)$). Rule $(\mathcal{E}13)$ caters to the case of definition calls with no name substitutions. Rule $(\mathcal{E}14)$ removes a silent guard from a process, as this has no effect on the interpretation, and resumes interpreting the residual process. Finally, rule $(\mathcal{E}15)$ resolves a name equality comparison process using again the $\overset{\phi_{\mathcal{E}}}{\sim}$ relation based on the current value of the $\phi_{\mathcal{E}}$ environment.

The following soundness theorem states that name substitutions in the structural operational semantics are captured in the non-standard semantics.

Theorem 2.1 (Soundness of the non-standard name-substitution semantics) $\forall P, Q,$ $x, y : P \overset{\mu}{\longrightarrow}{}^{*} Q[x/y] \Rightarrow x \in \phi_{\mathcal{E}}'(y)$ where, $\phi_{\mathcal{E}}' = \mathcal{E}[\![P]\!] \rho \phi_{\mathcal{E}}$

Proof sketch. The proof is by induction on the rules of the structural operational semantics in Fig. 2.1. The most interesting cases are rules (15) and (16), where we need to show that if a process, P, exhibits a transition, $P \overset{\overline{x}\langle y \rangle}{\longrightarrow} P'$ or $P \overset{\overline{x}\langle \rangle}{\longrightarrow} P'$, then this will eventually yield a process, $\overline{x}\langle y \rangle.P'' \in \rho$ or $\overline{x}\langle \rangle.P'' \in \rho$, during the non-standard interpretation. The same can be shown for $Q \overset{x(z)}{\longrightarrow} Q'$, $Q \overset{x()}{\longrightarrow} Q'$ and $R \overset{\overline{x}\langle y \rangle}{\longrightarrow} R'$. From rule $(\mathcal{E}3)$, we can then show that $P \mid Q$ and $R \mid Q$ will capture substitutions in the $\phi_{\mathcal{E}}$ environment in each case. Clock-ticking of rule (11) in the structural operational semantics is matched by the option of no communication and no synchronisation in rules $(\mathcal{E}3)$ and $(\mathcal{E}4)$ in the non-standard semantics. $\qquad\square$

The above theorem is important as it shows that the non-standard name-substitution semantics is sound with respect to the standard structural operational interpretation, and hence no irregular meaning is introduced in the non-standard semantics that cannot be justified in the standard semantics.

2.5 Examples

Let's consider a few examples of simple processes and their non-standard meaning:

- $P1 = ((\nu z)\overline{x}\langle z\rangle.0) \mid \texttt{timer}^{t+1}(x(y).0, 0)$
  $\mathcal{E}(\![P1]\!) \{\!\}\, \phi_{\mathcal{E}0} = \phi_{\mathcal{E}}[y_1 \mapsto \{z_1\}]$
- $P2 = (!(\nu z)\overline{x}\langle z\rangle.0) \mid \texttt{timer}^{t+1}(x(y).0, 0)$
  $\mathcal{E}(\![P2]\!) \{\!\}\, \phi_{\mathcal{E}0} = \phi_{\mathcal{E}}[y_1 \mapsto \{z_1, z_2, z_3, \ldots, z_\infty\}]$
- $P3 = (!(\nu z)\overline{x}\langle z\rangle.0) \mid (!\texttt{timer}^{t+1}(x(y).0, 0))$
  $\mathcal{E}(\![P3]\!) \{\!\}\, \phi_{\mathcal{E}0} = \phi_{\mathcal{E}}[y_1 \mapsto \{z_1, z_2, z_3, \ldots, z_\infty\}, y_2 \mapsto \{z_1, z_2, z_3, \ldots, z_\infty\}, y_3 \mapsto \{z_1, z_2, z_3, \ldots, z_\infty\}, \ldots, y_\infty \mapsto \{z_1, z_2, z_3, \ldots, z_\infty\}]$
- $P4 = ((\nu z)\overline{x}\langle z\rangle.0) + \texttt{timer}^{t+1}(x(y).0, 0)$
  $\mathcal{E}(\![P4]\!) \{\!\}\, \phi_{\mathcal{E}0} = \phi_{\mathcal{E}0}$
- $P5 = !(((\nu z)\overline{x}\langle z\rangle.0) + \texttt{timer}^{t+1}(x(y).0, 0))$
  $\mathcal{E}(\![P5]\!) \{\!\}\, \phi_{\mathcal{E}0} = \phi_{\mathcal{E}}[y_1 \mapsto \{z_1, z_2, z_3, \ldots, z_\infty\}, y_2 \mapsto \{z_1, z_2, z_3, \ldots, z_\infty\}, y_3 \mapsto \{z_1, z_2, z_3, \ldots, z_\infty\}, \ldots, y_\infty \mapsto \{z_1, z_2, z_3, \ldots, z_\infty\}]$
- $P6 = ((\nu z)\overline{x}\langle z\rangle.0) \mid \texttt{timer}^0(x(y).0, 0)$
  $\mathcal{E}(\![P6]\!) \{\!\}\, \phi_{\mathcal{E}0} = \phi_{\mathcal{E}0}$
- $P7 = ((\nu z)\overline{x}\langle z\rangle.0) \mid \texttt{timer}^0(x(y).0, \texttt{timer}^{t+1}(x(u).0, 0))$
  $\mathcal{E}(\![P7]\!) \{\!\}\, \phi_{\mathcal{E}0} = \phi_{\mathcal{E}}[u_1 \mapsto \{z_1\}]$
- $P8 = ((\nu z)\overline{x}\langle z\rangle.0) \mid \texttt{timer}^{t+1}(x(y).0, \texttt{timer}^{t'+1}(x(u).0, 0))$
  $\mathcal{E}(\![P8]\!) \{\!\}\, \phi_{\mathcal{E}0} = \phi_{\mathcal{E}}[y_1 \mapsto \{z_1\}, u_1 \mapsto \{z_1\}]$
- $P9 = ((\nu z)\overline{x}\langle z\rangle.0) \mid (\texttt{timer}^{t+1}(x(y).0, 0) + \texttt{timer}^{t'+1}(x(u).0, 0))$
  $\mathcal{E}(\![P9]\!) \{\!\}\, \phi_{\mathcal{E}0} = \phi_{\mathcal{E}}[y_1 \mapsto \{z_1\}, u_1 \mapsto \{z_1\}]$
- $P10 = (!\,\overline{x}\langle z\rangle.0) \mid \texttt{timer}^{t+1}(x(y).0, 0)$
  $\mathcal{E}(\![P10]\!) \{\!\}\, \phi_{\mathcal{E}0} = \phi_{\mathcal{E}}[y_1 \mapsto \{z\}]$

References

1. R. Milner, *Communicating and Mobile Systems: The Pi-Calculus* (Cambridge University Press, Cambridge, 1999)
2. R. Milner, J. Parrow, D. Walker, Inf. Comput. **100**(1), 1 (1992)
3. R. Milner, The polyadic π-calculus: a tutorial. Technical report ECS-LFCS-91-180, Laboratory for Foundations of Computer Science, Department of Computer Science, University of Edinburgh, UK (1991)
4. G. Boudol, Asynchrony and the π-calculus. Technical report 1702, INRIA-Sophia Antipolis, Sophia Antipolis, France (1992)

5. K. Honda, M. Tokoro, in *Proceedings of the ECOOP'91 Workshop on Object-Based Concurrent Computing*, eds. by M. Tokoro, O. Nierstrasz, P. Wegner. Lecture Notes in Computer Science, vol. 612 (Springer, Geneva, Switzerland, 1992), pp. 21–51
6. J.W. Backus, in *Information Processing, Proceedings of the 1st International Conference on Information Processing, UNESCO, Paris 15-20 June 1959* (UNESCO (Paris), 1959), pp. 125–131
7. J.W. Backus, F.L. Bauer, J. Green, C. Katz, J.L. McCarthy, A.J. Perlis, H. Rutishauser, K. Samelson, B. Vauquois, J.H. Wegstein, A. van Wijngaarden, M. Woodger, P. Naur, Commun. ACM **6**(1), 1 (1963)
8. B. Aziz, in *9th International Conference on Availability, Reliability and Security (ARES 2014), Fribourg, Switzerland* (IEEE, 2014)
9. R. Milner, Lect. Notes Comput. Sci. **92** (1980)

Chapter 3
Formal Analysis by Abstract Interpretation

3.1 Introduction

After defining a non-standard denotational semantics for our formal language in the
previous chapter, we use these semantics as the basis for constructing an abstract
interpretation from which we can perform a static analysis of specifications written
in this formal language. This involves the introduction of a suitable abstraction (or
approximation) that restricts the semantic domain holding the meaning of a process,
to a finite size, thereby allowing the termination of any least fixed-point calculations
over that semantic domain. This abstraction is required to be safe with respect to the
concrete non-standard semantics; a term substitution occurring in the concrete (non-
standard) semantics will necessarily be captured in an abstract manner by the abstract
semantics. However, due to the imprecise nature of abstractions, false positives may
appear in the final results. For example, in the process

$$\overline{x}\langle s\rangle \mid xx(y) \mid \overline{z}\langle w\rangle \mid zz(u)$$

if, $x = xx = t_1$ and $z = zz = t_2$, then we have in the standard semantics that s can
only substitute y, and w can only substitute u. However, abstracting the above values,
such that $x = xx \in \{t_1, t_2\}$ and $z = zz \in \{t_1, t_2\}$, then s may substitute y or u, and
similarly, w may substitute y or u. This is because the abstraction introduces the
scenario where $x = zz$ and $z = xx$, which is impossible in the concrete semantics.

Finally, this chapter gives specifications of a couple of examples of simple protocols, where we apply to them the abstract interpretation we define, in order to capture
the property of name substitutions.

© The Author(s), under exclusive license to Springer Nature Switzerland AG 2022   43
B. Aziz, *Formal Analysis by Abstract Interpretation*, SpringerBriefs in Applied
Sciences and Technology, https://doi.org/10.1007/978-3-030-91153-9_3

3.2   An Abstract Domain

To obtain a computable semantics, i.e. whose termination is guaranteed, a safe abstraction is required to remove the source of infinite growth in the semantic domain $D_\perp$. For example, consider the following system:

$$!(((\nu z)\overline{x}\langle z\rangle.\mathbf{0}) \mid \texttt{timer}^\infty(x(y).\mathbf{0}, \mathbf{0}))$$

As a result of the infinite communications within this system, its non-standard semantics will yield (as part of its meaning) an infinite $\phi_{\mathcal{A}}$ environment that maps every instance y_j to $\{z_j\}$, for every value of $j \in \mathbb{N}$.

Therefore, we need to introduce a safe approximation for the non-standard meaning of processes by introducing an approximation function, α_k, which limits the number of copies of bound names in a process generated as a result of infinite replication to k number of copies.

Definition 3.1 (*The α_k-approximation function*) Define $\alpha_k : N \to N^\sharp$ as follows, where $N^\sharp = (N \backslash \{x_i \mid i > k\})$:

$$\forall x \in N : \alpha_k(x) = \begin{cases} x_k & \text{if } x = x_i \ \wedge \ i > k \\ x & \text{otherwise} \end{cases}$$

□

We write $\alpha_k(\{x, x', \ldots\})$ to mean $\{\alpha_k(x), \alpha_k(x'), \ldots\}$. The α_k approximation function leads naturally to the appearance of an abstract name-substitution environment, $\phi_{\mathcal{A}} : N^\sharp \to \wp(N^\sharp)$, and an abstract semantic domain, $D_\perp^\sharp$, with the following ordering:

$$\forall \phi_{\mathcal{A}1}, \phi_{\mathcal{A}2} \in D_\perp^\sharp : \ \phi_{\mathcal{A}1} \sqsubseteq_{D_\perp^\sharp} \phi_{\mathcal{A}2} \ \Leftrightarrow \ \forall x \in N^\sharp : \ \phi_{\mathcal{A}1}(x) \subseteq \phi_{\mathcal{A}2}(x)$$

The bottom element, $\perp_{D_\perp^\sharp}$, is the null environment, $\phi_{\mathcal{A}0}$, mapping every abstract name to the empty set. Abstract environment union $\cup_{\phi_{\mathcal{A}}}$ is defined in a similar fashion to its precise counterpart using set union:

$$\forall \phi_{\mathcal{A}1}, \phi_{\mathcal{A}2} \in D_\perp^\sharp, x \in N^\sharp : (\phi_{\mathcal{A}1} \cup_{\phi_{\mathcal{A}}} \phi_{\mathcal{A}2})(x) = \phi_{\mathcal{A}1}(x) \cup \phi_{\mathcal{A}2}(x)$$

3.3   An Abstract Interpretation Function

Based on the abstract semantic domain, $D_\perp^\sharp$, defined in the previous section, we can now give an abstract interpretation for processes in our language using the abstract semantic function, $\mathcal{A}[\![P]\!] \ \rho \ \phi_{\mathcal{A}} \in D_\perp^\sharp$. The syntax-directed rules for this function are given in Fig. 3.1.

These rules are quite identical to those given for the non-standard semantics of Fig. 2.2, except that they constrain the number of copies of bound names (i.e. input

$(\mathcal{A}1)$ $\mathcal{A}([\overline{x}\langle y\rangle.P]) \, \rho \, \phi_{\mathcal{A}}$ $= \phi_{\mathcal{A}}$

$(\mathcal{A}2)$ $\mathcal{A}([\overline{x}\langle\rangle.P]) \, \rho \, \phi_{\mathcal{A}}$ $= \phi_{\mathcal{A}}$

$(\mathcal{A}3)$ $\mathcal{A}([\mathtt{timer}^{t+1}(x(y).P,Q)]) \, \rho \, \phi_{\mathcal{A}} =$

$\qquad (\quad \bigcup_{\phi_{\mathcal{A}}} \underset{R\in\rho}{\mathcal{R}}([(\uplus \{|R|\}) \uplus \{|P|\} \uplus \{|P'|\}]) \, \phi_{\mathcal{A}}[\alpha_k(y) \mapsto \phi_{\mathcal{A}}(\alpha_k(y)) \cup \{\alpha_k(z)\}])$

$\qquad \underset{\overline{x'}\langle z\rangle.P'\in\rho:\, x \overset{\phi_{\mathcal{A}}}{\sim} x'}{}$

$\qquad \cup_{\phi_{\mathcal{A}}} \underset{R\in\rho}{\mathcal{R}}([(\uplus \{|R|\}) \uplus \{|\mathtt{timer}^{t}(x(y).P,Q)|\}]) \, \phi_{\mathcal{A}}$

$(\mathcal{A}4)$ $\mathcal{A}([\mathtt{timer}^{t+1}(x().P,Q)]) \, \rho \, \phi_{\mathcal{A}} \quad =$

$\qquad (\quad \bigcup_{\phi_{\mathcal{A}}} \underset{R\in\rho}{\mathcal{R}}([(\uplus \{|R|\}) \uplus \{|P|\} \uplus \{|P'|\}]) \, \phi_{\mathcal{A}}) \cup_{\phi_{\mathcal{A}}}$

$\qquad \underset{\overline{x'}\langle\rangle.P'\in\rho:\, x \overset{\phi_{\mathcal{A}}}{\sim} x'}{}$

$\qquad \mathcal{R}([(\uplus \{|R|\}) \uplus \{|\mathtt{timer}^{t}(x(y).P,Q)|\}]) \, \phi_{\mathcal{A}}$

$\qquad \underset{R\in\rho}{}$

$(\mathcal{A}5)$ $\mathcal{A}([\mathtt{timer}^{0}(x(y).P,Q)]) \, \rho \, \phi_{\mathcal{A}} \;= \mathcal{R}([\{|Q|\} \uplus \rho]) \, \phi_{\mathcal{A}}$

$(\mathcal{A}6)$ $\mathcal{A}([\mathtt{timer}^{0}(x().P,Q)]) \, \rho \, \phi_{\mathcal{A}} \;= \mathcal{R}([\{|Q|\} \uplus \rho]) \, \phi_{\mathcal{A}}$

$(\mathcal{A}7)$ $\mathcal{A}([P \mid Q]) \, \rho \, \phi_{\mathcal{A}} \qquad\qquad = \mathcal{R}([\{|P|\} \uplus \{|Q|\} \uplus \rho]) \, \phi_{\mathcal{A}}$

$(\mathcal{A}8)$ $\mathcal{A}([!P]) \, \rho \, \phi_{\mathcal{A}} \qquad\qquad\quad = snd(fix \, \mathcal{F}(0, \perp_{D^{\sharp}}))$

$\qquad$ where, $\mathcal{F} = \lambda f \, \lambda(j, \phi).$

$$f \; (if \phi = \mathcal{R}([(\overset{j}{\underset{i=0}{\uplus}} \{|(P)\sigma|\}) \uplus \rho]) \, \phi_{\mathcal{A}} \; then$$

$$j, \phi \; else \; (j+1), (\mathcal{R}([(\overset{j}{\underset{i=0}{\uplus}} \{|(P)\sigma|\}) \uplus \rho]) \, \phi_{\mathcal{A}}))$$

$\qquad$ and $\sigma = [bn_i(P)/bn(P)], \; bn_i(P) = \{x_i \mid x \in bn(P)\}$

$(\mathcal{A}9)$ $\mathcal{A}([(\nu n)P]) \, \rho \, \phi_{\mathcal{A}} \qquad\qquad = \mathcal{R}([\{|P|\} \uplus \rho]) \, \phi_{\mathcal{A}}$

$(\mathcal{A}10)$ $\mathcal{A}([0]) \, \rho \, \phi_{\mathcal{A}} \qquad\qquad\quad = \phi_{\mathcal{A}}$

$(\mathcal{A}11)$ $\mathcal{A}([P+Q]) \, \rho \, \phi_{\mathcal{A}} \qquad\qquad = \mathcal{A}([P]) \, \rho \, \phi_{\mathcal{A}} \; \cup_{\phi_{\mathcal{A}}} \; \mathcal{A}([Q]) \, \rho \, \phi_{\mathcal{A}}$

$(\mathcal{A}12)$ $\mathcal{A}([A(x)]) \, \rho \, \phi_{\mathcal{A}} \qquad\quad\; = \mathcal{A}([P[x/y]]) \, \rho \, \phi_{\mathcal{A}} \; where, \; A(y) \overset{def}{=} P$

$(\mathcal{A}13)$ $\mathcal{A}([A]) \, \rho \, \phi_{\mathcal{A}} \qquad\qquad\quad = \mathcal{A}([P]) \, \rho \, \phi_{\mathcal{A}} \; where, \; A() \overset{def}{=} P$

$(\mathcal{A}14)$ $\mathcal{A}([\tau.P]) \, \rho \, \phi_{\mathcal{A}} \qquad\qquad = \mathcal{A}([P]) \, \rho \, \phi_{\mathcal{A}}$

$(\mathcal{A}15)$ $\mathcal{A}([P[x=y]Q]) \, \rho \, \phi_{\mathcal{A}} \qquad = \begin{cases} \mathcal{A}([P]) \, \rho \, \phi_{\mathcal{A}} & if \, x \overset{\phi_{\mathcal{A}}}{\sim} y \\ \mathcal{A}([Q]) \, \rho \, \phi_{\mathcal{A}} & otherwise \end{cases}$

$(\mathcal{R}0)$ $\mathcal{R}([\rho]) \, \phi_{\mathcal{A}} \qquad\qquad\qquad = \underset{P\in\rho}{\bigcup_{\phi_{\mathcal{A}}}} \mathcal{A}([P]) \, (\rho \backslash \{|P|\}) \, \phi_{\mathcal{A}}$

Fig. 3.1 The definition of the abstract interpretation function, $\mathcal{A}([P]) \, \rho \, \phi_{\mathcal{A}}$

parameters and possibly bound messages) to a maximum limit of k copies using the approximation function, α_k. This approximation is only needed for rule $(\mathcal{A}3)$, where message-passing communication takes place and therefore, it will affect the value of the input parameter in the $\phi_{\mathcal{A}}$ environment. The other difference in this abstract semantics is the use of the abstract name equivalence relation, $\overset{\phi_{\mathcal{A}}}{\sim}$, defined as follows:

$$x \overset{\phi_{\mathcal{A}}}{\sim} y \; \Leftrightarrow \; (\phi_{\mathcal{A}}(x) \cap \phi_{\mathcal{A}}(y) \neq \emptyset) \vee (x = y)$$

used in rules $(\mathcal{A}3)$, $(\mathcal{A}4)$ and $(\mathcal{A}15)$.

Next, we need to show that this abstract interpretation indeed terminates, in order for it to be more useful than the non-standard semantics.

Theorem 3.1 (Termination of the Abstract Semantics) *For any process, P, computing $\mathcal{A}([P]) \, \rho \perp_{D^{\sharp}_{\perp}}$ terminates.*

Proof sketch. The proof relies on two requirements: First, to show that $D_\perp^\sharp$ is finite. This is true from the definition of α_k as given in Definition 3.1. The second is to show that the abstract meaning of a process is monotonic with respect to the number of copies of a replicated process, more formally that

$$\mathcal{R}((\biguplus_{i=0}^{j} \{\!|P|\!\}) \uplus \rho) \phi_\mathcal{A} \sqsubseteq_{D_\perp^\sharp} \mathcal{R}((\biguplus_{i=0}^{j+1} \{\!|P|\!\}) \uplus \rho) \phi_\mathcal{A}$$

The second requirement is proven by showing that the extra copy of P can "only" induce more communications, and not fewer ones. As a result, and given the fixed-point calculation of rule $(\mathcal{A}8)$, $\mathcal{A}(\![P]\!) \rho \perp_{D_\perp^\sharp}$ terminates. $\square$

The next property we need to prove that holds for the abstract semantics is that it is safe with regards to the non-standard semantics. Informally, this means that no communications (i.e. name substitutions) captured in the non-standard semantics will be missed in the abstract semantics. In order to be able to prove the safety of the abstract semantics with respect to the non-standard semantics, we first prove the safety of the union of abstract environments, $\cup_{\phi_\mathcal{A}}$, operator, more formally as follows.

Lemma 3.1 (Safety of the $\cup_{\phi_\mathcal{A}}$ operator)
$$\forall i \in \{0\dots n\}, n \in \mathbb{N}, \phi_{\mathcal{E}i} \in D_\perp, \phi_{\mathcal{A}i} \in D_\perp^\sharp, \phi = \bigcup_{\substack{\phi_\mathcal{E} \\ i=0\dots n}} \phi_{\mathcal{E}i}, \phi^\sharp = \bigcup_{\substack{\phi_\mathcal{A} \\ i=0\dots n}} \phi_{\mathcal{A}i} :$$
$$(\exists x, y \in \mathcal{N} : \ (y \in \phi_{\mathcal{E}i}(x) \Rightarrow \alpha_k(y) \in \phi_{\mathcal{A}i}(\alpha_k(x)))) \Rightarrow$$
$$(\exists x, y \in \mathcal{N} : \ (y \in \phi(x) \Rightarrow \alpha_k(y) \in \phi^\sharp(\alpha_k(x))))$$

Proof The proof proceeds by mathematical induction on sets of environments.
The base case: $n = 0$
From $(\phi = \bigcup_{\substack{\phi_\mathcal{E} \\ i=0\dots0}} \phi_{\mathcal{E}i} = \phi_{\mathcal{E}0})$ and $(\phi^\sharp = \bigcup_{\substack{\phi_\mathcal{A} \\ i=0\dots0}} \phi_{\mathcal{A}i} = \phi_{\mathcal{A}0})$, we have that
$$(\exists x, y \in \mathcal{N} : \ (y \in \phi_{\mathcal{E}i}(x) \Rightarrow \alpha_k(y) \in \phi_{\mathcal{A}i}(\alpha_k(x)))) \Rightarrow$$
$$(\exists x, y \in \mathcal{N} : \ (y \in \phi(x) \Rightarrow \alpha_k(y) \in \phi^\sharp(\alpha_k(x))))$$
since $\phi_{\mathcal{E}0} = \phi = \{\}$ and $\phi_{\mathcal{A}0} = \phi^\sharp = \{\}$.
The induction step:
First, assume the hypothesis is true for a union of n number of environments:
$$\forall i \in \{0\dots n\}, n \in \mathbb{N}, \phi_{\mathcal{E}i} \in D_\perp, \phi_{\mathcal{A}i} \in D_\perp^\sharp, \phi = \bigcup_{\substack{\phi_\mathcal{E} \\ i=0\dots n}} \phi_{\mathcal{E}i}, \phi^\sharp = \bigcup_{\substack{\phi_\mathcal{A} \\ i=0\dots n}} \phi_{\mathcal{A}i} :$$
$$(\exists x, y \in \mathcal{N} : \ (y \in \phi_{\mathcal{E}i}(x) \Rightarrow \alpha_k(y) \in \phi_{\mathcal{A}i}(\alpha_k(x)))) \Rightarrow$$
$$(\exists x, y \in \mathcal{N} : \ (y \in \phi(x) \Rightarrow \alpha_k(y) \in \phi^\sharp(\alpha_k(x))))$$
Proving the induction step demands that the above statement is also true for a union of $n + 1$ number of environments. In other words, assuming that we have $\phi = \bigcup_{\substack{\phi_\mathcal{E} \\ i=0\dots n}} \phi_{\mathcal{E}i}$
and $\phi^\sharp = \bigcup_{\substack{\phi_\mathcal{A} \\ i=0\dots n}} \phi_{\mathcal{A}i}$, as above, then,
$$\bigcup_{\substack{\phi_\mathcal{E} \\ i=0\dots n+1}} \phi_{\mathcal{E}i} = \phi \cup_{\phi_\mathcal{E}} \phi_{\mathcal{E}n+1} \text{ and } \bigcup_{\substack{\phi_\mathcal{A} \\ i=0\dots n+1}} \phi_{\mathcal{A}i} = \phi^\sharp \cup_{\phi_\mathcal{A}} \phi_{\mathcal{A}n+1}$$
From the definitions of $\cup_{\phi_\mathcal{E}}$ and $\cup_{\phi_\mathcal{A}}$, we have also

$$(\phi \cup_{\phi_\mathcal{E}} \phi_{\mathcal{E}n+1})(x) = \phi(x) \cup \phi_{\mathcal{E}n+1}(x) \text{ and}$$
$$(\phi^\sharp \cup_{\phi_\mathcal{A}} \phi_{\mathcal{A}n+1})(x) = \phi^\sharp(x) \cup \phi_{\mathcal{A}n+1}(x)$$

which leads to the result that $\forall y \in \phi(x), y' \in \phi_{\mathcal{E}_{n+1}}(x) \Rightarrow y, y' \in (\phi(x) \cup \phi_{\mathcal{E}n+1}(x))$ and that $\forall y \in \phi^{\sharp}(x), y' \in \phi_{\mathcal{A}_{n+1}}(x) \Rightarrow y, y' \in (\phi^{\sharp}(x) \cup \phi_{\mathcal{A}n+1}(x))$. However, by the inductive hypothesis, we have that

$$y \in \phi(x) \Rightarrow \alpha_k(y) \in \phi^{\sharp}(\alpha_k(x))$$

and with the assumption that the newly added environments are also safe, that is,

$$y \in \phi_{\mathcal{E}n+1}(x) \Rightarrow \alpha_k(y) \in \phi_{\mathcal{A}n+1}(\alpha_k(x))$$

Then by the properties of set union, it is possible to arrive at the following final result:

$$y \in (\phi(x) \cup \phi_{\mathcal{E}n+1}(x))$$
$$\Rightarrow \alpha_k(y) \in (\phi^{\sharp}(\alpha_k(x)) \cup \phi_{\mathcal{A}n+1}(\alpha_k(x))) \qquad \square$$

The lemma hence states that for any set of concrete environments, $\{\phi_{\mathcal{E}1}, \ldots, \phi_{\mathcal{E}n}\}$, and their abstract counterparts, $\{\phi_{\mathcal{A}1}, \ldots, \phi_{\mathcal{A}n}\}$, which are related by the abstraction function, α_k, the unions, $\phi = \bigcup_{\phi_{\mathcal{E}}\ i=0...n} \phi_{\mathcal{E}i}$ and $\phi^{\sharp} = \bigcup_{\phi_{\mathcal{A}}\ i=0...n} \phi_{\mathcal{A}i}$, will maintain this relation. Intuitively, the relation states that a name, y, captured by a $\phi_{\mathcal{E}}$ environment in the concrete semantics, will appear as $\alpha_k(y)$ in the corresponding $\phi_{\mathcal{A}}$ environment.

From this lemma, the safety of the abstract semantics can now be established formally by the following theorem.

Theorem 3.2 (Safety of the abstract semantics)

$\forall P, \rho, \phi_{\mathcal{E}}, \phi_{\mathcal{A}}, \mathcal{E}[\![P]\!] \rho \phi_{\mathcal{E}} = \phi'_{\mathcal{E}}, \mathcal{A}[\![P]\!] \rho \phi_{\mathcal{A}} = \phi'_{\mathcal{A}}:$
$(\exists x, y \in \mathcal{N} : y \in \phi_{\mathcal{E}}(x) \Rightarrow \alpha_k(y) \in \phi_{\mathcal{A}}(\alpha_k(x))) \Rightarrow$
$(\exists x, y \in \mathcal{N} : y \in \phi'_{\mathcal{E}}(x) \Rightarrow \alpha_k(y) \in \phi'_{\mathcal{A}}(\alpha_k(x)))$

Proof sketch. The proof proceeds by structural induction on the structure of processes, P. We start first with the base cases, where $\phi_{\mathcal{E}} = \phi'_{\mathcal{E}}$ and $\phi_{\mathcal{A}} = \phi'_{\mathcal{A}}$. These include rules ($\mathcal{A}$1), ($\mathcal{A}$2) and ($\mathcal{A}$10). For all of these cases, our condition becomes

$(\exists x, y \in \mathcal{N} : y \in \phi_{\mathcal{E}}(x) \Rightarrow \alpha_k(y) \in \phi_{\mathcal{A}}(\alpha_k(x))) \Rightarrow$
$(\exists x, y \in \mathcal{N} : y \in \phi_{\mathcal{E}}(x) \Rightarrow \alpha_k(y) \in \phi_{\mathcal{A}}(\alpha_k(x)))$

which will always be true from the definition of a logical implication. Next, we consider the case of rule ($\mathcal{A}$3). For this case, we can prove that the resulting environment $\phi'_{\mathcal{A}}$ is safe since it relies in its building on the operator, $\cup_{\phi_{\mathcal{A}}}$. And since $\cup_{\phi_{\mathcal{A}}}$ is a safe operator (Lemma 3.1), then the right-hand side of the implication will also be true. The same argument can be used for all the other rules that use the $\cup_{\phi_{\mathcal{A}}}$ operator, such as rules ($\mathcal{A}$4), ($\mathcal{A}$11) and ($\mathcal{R}$0). For rules that change the structure of the ρ environment of parallel processes, the safety condition is passed on to the new parallel processes to the right-hand side of the equivalence in the rule. This applies to rules such as ($\mathcal{A}$5), ($\mathcal{A}$6), ($\mathcal{A}$7) and ($\mathcal{A}$9). The same argument can be used for the cases of the rules that merely pass on the analysis to the body of the process but not modify ρ, such as rules ($\mathcal{A}$12), ($\mathcal{A}$13), ($\mathcal{A}$14) and ($\mathcal{A}$15).

Finally, the case of replication in rule ($\mathcal{A}$8) requires delicate consideration here as it combines mathematical induction with structural induction. We outline this last case in more detail. First, we demonstrate that $\mathcal{F}_{\mathcal{A}}(-1, \perp_{D^\sharp})$ is safe with respect to $\mathcal{F}_{\mathcal{E}}(-1, \perp_D)$ (base case) and that if $\mathcal{F}_{\mathcal{A}}(n, \perp_{D^\sharp})$ is safe with respect to $\mathcal{F}_{\mathcal{E}}(n, , \perp_D)$, then $\mathcal{F}_{\mathcal{A}}(n + 1, \perp_{D^\sharp})$ will also be safe with respect to $\mathcal{F}_{\mathcal{E}}(n + 1, \perp_D)$ (inductive step).

The base case: $n = -1$

$$\mathcal{F}_{\mathcal{A}}(-1, \perp_{D^\sharp}) = \; let \; \phi_1 = \perp_{D^\sharp} \; in$$
$$let \; \phi_2 = \mathcal{A}[\![\, \mathbf{0} \,]\!] \; \rho \; \phi_{\mathcal{A}} \; in$$
$$if \; \phi_1 = \phi_2 \; then \; \phi_1 \; else \; \mathcal{F}_{\mathcal{A}}(0, \phi_2)$$
$$\mathcal{F}_{\mathcal{E}}(-1, \perp_D) = \; let \; v_1 = \perp_D \; in$$
$$let \; v_2 = \mathcal{E}[\![\, \mathbf{0} \,]\!] \; \rho \; \phi_{\mathcal{E}} \; in$$
$$if \; v_1 = v_2 \; then \; v_1 \; else \; \mathcal{F}_{\mathcal{E}}(0, v_2)$$

Note first in the above that $\mathcal{F}_{\mathcal{A}}(-1, \perp_{D^\sharp})$ means no copies of P are spawned, whereas $\mathcal{F}_{\mathcal{A}}(0, \phi_2)$ means that one copy is spawned (numbered copy 0). The same is true for $\mathcal{F}_{\mathcal{E}}(-1, \perp_D)$ and $\mathcal{F}_{\mathcal{E}}(0, v_2)$, respectively. From the above two calculations, we have that $\perp_{D^\sharp} = \phi_{\mathcal{A}0}$ is a safe abstraction of $\perp_D = \phi_{\mathcal{E}0}$ since $\forall x \in dom(\phi_{\mathcal{E}0}) : \phi_{\mathcal{E}0}(x) = \phi_{\mathcal{A}0}(x) = \{\}$. Also, since we have proved the case for $P = \mathbf{0}$ is safe earlier, then we have that

$$\forall P, \rho, \phi_{\mathcal{E}}, \phi_{\mathcal{A}}, \mathcal{E}[\![P]\!] \; \rho \; \phi_{\mathcal{E}} = \phi'_{\mathcal{E}}, \mathcal{A}[\![P]\!] \; \rho \; \phi_{\mathcal{A}} = \phi'_{\mathcal{A}} :$$
$$(\exists x, y \in \mathcal{N} : y \in \phi_{\mathcal{E}}(x) \; \Rightarrow \; \alpha_k(y) \in \phi_{\mathcal{A}}(\alpha_k(x))) \; \Rightarrow$$
$$(\exists x, y \in \mathcal{N} : y \in \phi'_{\mathcal{E}}(x) \; \Rightarrow \; \alpha_k(y) \in \phi'_{\mathcal{A}}(\alpha_k(x)))$$

The induction step:

We start by assuming first the inductive hypothesis, i.e. for the case of n copies of the replicated process:

$$\mathcal{F}_{\mathcal{E}}(n, \phi_{\mathcal{E}}) = \phi'_{\mathcal{E}} \wedge \mathcal{F}_{\mathcal{A}}(n, \phi_{\mathcal{A}}) = \phi'_{\mathcal{A}} :$$
$$(\exists x, y \in \mathcal{N} : y \in \phi_{\mathcal{E}}(x) \; \Rightarrow \; \alpha_k(y) \in \phi_{\mathcal{A}}(\alpha_k(x))) \; \Rightarrow$$
$$(\exists x, y \in \mathcal{N} : y \in \phi'_{\mathcal{E}}(x) \; \Rightarrow \; \alpha_k(y) \in \phi'_{\mathcal{A}}(\alpha_k(x)))$$

Now, we prove that the same safety relation holds for $\mathcal{F}_{\mathcal{A}}(n + 1, \phi_{\mathcal{A}})$ with respect to $\mathcal{F}_{\mathcal{E}}(n + 1, \phi_{\mathcal{E}})$, as follows:

$$\mathcal{F}_{\mathcal{A}}(n + 1, \phi_{\mathcal{A}}) = \; let \; \phi_1 = \mathcal{A}[\![\prod_{i=0}^{n+1} (P)\sigma \,]\!] \; \rho \; \phi_{\mathcal{A}} \; in$$

$$let \; \phi_2 = \mathcal{A}[\![\prod_{i=0}^{n+2} (P)\sigma \,]\!] \; \rho \; \phi_{\mathcal{A}} \; in$$

$$if \; \phi_1 = \phi_2 \; then \; \phi_1 \; else \; \mathcal{F}_{\mathcal{A}}(n + 2, \phi_2)$$

$$\mathcal{F}_{\mathcal{E}}(n + 1, \phi_{\mathcal{E}}) = \; let \; v_1 = \mathcal{E}[\![\prod_{i=0}^{n+1} (P)\sigma \,]\!] \; \rho \; \phi_{\mathcal{E}} \; in$$

$$let \; v_2 = \mathcal{E}[\![\prod_{i=0}^{n+2} (P)\sigma \,]\!] \; \rho \; \phi_{\mathcal{E}} \; in$$

$$if \; v_1 = v_2 \; then \; v_1 \; else \; \mathcal{F}_{\mathcal{E}}(n + 2, v_2)$$

From the induction hypothesis, we know that

$$\mathcal{E}[\![\prod_{i=0}^{n+1} (P)\sigma \,]\!] \; \rho \; \phi_{\mathcal{E}} = \phi'_{\mathcal{E}} \wedge \mathcal{A}[\![\prod_{i=0}^{n+1} (P)\sigma \,]\!] \; \rho \; \phi_{\mathcal{A}} = \phi'_{\mathcal{A}} :$$
$$(\exists x, y \in \mathcal{N} : y \in \phi_{\mathcal{E}}(x) \; \Rightarrow \; \alpha_k(y) \in \phi_{\mathcal{A}}(\alpha_k(x))) \; \Rightarrow$$
$$(\exists x, y \in \mathcal{N} : y \in \phi'_{\mathcal{E}}(x) \; \Rightarrow \; \alpha_k(y) \in \phi'_{\mathcal{A}}(\alpha_k(x)))$$

The remaining case involves $\mathcal{E}[\![\prod_{i=0}^{n+2}(P)\sigma]\!] \rho \phi_{\mathcal{E}}$ and $\mathcal{A}[\![\prod_{i=0}^{n+2}(P)\sigma]\!] \rho \phi_{\mathcal{A}}$. We can rewrite these as follows:

$$\mathcal{E}[\![\prod_{i=0}^{n+2}(P)\sigma]\!] \rho \phi_{\mathcal{E}} = \mathcal{E}[\![(\prod_{i=0}^{n+1}(P)\sigma) \mid (\prod_{i=n+2}^{n+2}(P)\sigma)]\!] \rho \phi_{\mathcal{E}}$$

$$\mathcal{A}[\![\prod_{i=0}^{n+2}(P)\sigma]\!] \rho \phi_{\mathcal{A}} = \mathcal{A}[\![(\prod_{i=0}^{n+1}(P)\sigma) \mid (\prod_{i=n+2}^{n+2}(P)\sigma)]\!] \rho \phi_{\mathcal{A}}$$

Moreover, from the antecedent, we know that
$\mathcal{E}[\![(P)\sigma_{n+2}]\!] \rho \phi_{\mathcal{E}} = \phi'_{\mathcal{E}} \wedge \mathcal{A}[\![(P)\sigma_{n+2}]\!] \rho \phi_{\mathcal{A}} = \phi'_{\mathcal{A}}$:
$(\exists x, y \in \mathcal{N} : y \in \phi_{\mathcal{E}}(x) \Rightarrow \alpha_k(y) \in \phi_{\mathcal{A}}(\alpha_k(x))) \Rightarrow$
$(\exists x, y \in \mathcal{N} : y \in \phi'_{\mathcal{E}}(x) \Rightarrow \alpha_k(y) \in \phi'_{\mathcal{A}}(\alpha_k(x)))$
And since we have already proven that the case for the parallel composition of the two processes $\mathcal{A}[\![P \mid Q]\!] \rho \phi_{\mathcal{A}}$ and $\mathcal{E}[\![P \mid Q]\!] \rho \phi_{\mathcal{E}}$ is safe, therefore, we can conclude that the inductive step is also safe:
$\mathcal{F}_{\mathcal{E}}(n+1, \phi_{\mathcal{E}}) = \phi'_{\mathcal{E}} \wedge \mathcal{F}_{\mathcal{A}}(n+1, \phi_{\mathcal{A}}) = \phi'_{\mathcal{A}}$:
$(\exists x, y \in \mathcal{N} : y \in \phi_{\mathcal{E}}(x) \Rightarrow \alpha_k(y) \in \phi_{\mathcal{A}}(\alpha_k(x))) \Rightarrow$
$(\exists x, y \in \mathcal{N} : y \in \phi'_{\mathcal{E}}(x) \Rightarrow \alpha_k(y) \in \phi'_{\mathcal{A}}(\alpha_k(x)))$
From both the base case and the induction step, we can conclude
$\mathcal{E}[\![!P]\!] \rho \phi_{\mathcal{E}} = \phi'_{\mathcal{E}} \wedge \mathcal{A}[\![!P]\!] \rho \phi_{\mathcal{A}} = \phi'_{\mathcal{A}}$:
$(\exists x, y \in \mathcal{N} : y \in \phi_{\mathcal{E}}(x) \Rightarrow \alpha_k(y) \in \phi_{\mathcal{A}}(\alpha_k(x))) \Rightarrow$
$(\exists x, y \in \mathcal{N} : y \in \phi'_{\mathcal{E}}(x) \Rightarrow \alpha_k(y) \in \phi'_{\mathcal{A}}(\alpha_k(x)))$ □

Informally, the safety theorem states that values present in the final environment resulting from the concrete non-standard semantics will always be present, as their abstract values, in the environment resulting from the abstract semantics.

3.3.1 Simple Example 1: An FTP Server

We consider here the first simple example to demonstrate the applicability of the abstract interpretation function we developed in the previous section, namely an example of a File Transfer Protocol (FTP) system. This system can be specified as follows:

$ftp \overset{def}{=}$ (! start (x).(ν login)(if x = start then (ν pwd)(Server | Client)
                                 else ((ν pwd)(Server) | I)

)
) | $\overline{start}\langle start^{t^1}\rangle.\overline{start}\langle start^{t^2}\rangle.\overline{start}\langle start^{t^3}\rangle.\overline{start}\langle I\_start^{t^4}\rangle$
$Server \overset{def}{=}$ (ν deal) login(z).if z = pwd then
    login(data).(
    $\overline{deal}\langle data^{t^5}\rangle$ | !deal(w).$\overline{login}\langle w^{t^6}\rangle$.login(u).if u = logout then 0
    else deal$\langle u^{t^7}\rangle$) else 0
$Client \overset{def}{=}$ (ν request) ($\overline{login}\langle pwd^{t^8}\rangle.\overline{login}\langle request^{t^9}\rangle$.login(res).$\overline{login}\langle logout^{t^{10}}\rangle$))

In this specification, the three *start* signals spawn three instances of the *Client/Server* system and the fourth *I_start* signal allows some intruder, *I*, to participate in a session with the server. Each instance of the client/server system shares a one-time password, *pwd*, that is not reusable. Additionally, communications between the server and the client or intruder processes are carried out over a secure login channel, *login* (which could be a Secure Socket Layer (SSL) connection). The client process performs a sequence of communications, sending the password and the request for data, and inputting the result from the server. After that, the client process logs out.

The server process accepts a password over the login channel after which it checks whether that password matches the current copy it shares with the client (intruder). If the password is correct, it waits for a data request and then *deals* with that request. For simplicity, the server just routes back the request over the session channel without altering it. The server process then waits for another input, after which it terminates if the new signal is *logout*. If not, it deals again with further requests sending the results over the login channel.

Often when analysing properties of systems, it is the case that the external context is taken into consideration, so as to achieve a more robust and modular analysis. This context is sometimes viewed as the larger and more powerful network, which is an arbitrary collection of intruder processes, I, assumed to be running concurrently with the system under analysis. The capabilities of this intruder include intercepting, reading and modifying any messages that travel over public channels as well as creating fresh messages, sending them to other processes and using them later for further communications. These capabilities vary depending on the network. For example, a local area network may be limited in its ability to communicate with computers external to the network as a result of its isolation from other networks by a firewall. It is also less able to create communication noise as a result of the high quality of service normally expected from local area networks. On the other hand, the Internet has a vast amount of information and is capable of communicating with virtually every computer that has some sort of connection to it (either directly or via a gateway).

For the case of the weakest intruder, its specification is $I \stackrel{\text{def}}{=} 0$. However, the model of I adopted in our framework draws on the lines of the most general attacker as given by Dolev and Yao in [1]. In the case of non-cryptographic systems, this model lacks the cryptography aspect and only captures the input/output capabilities of the Dolev–Yao model, as follows:

$$I \stackrel{\text{def}}{=} (\nu i) \, (\bar{i}\langle \kappa_{init} \rangle \mid !\, i(\kappa).(\prod_{\forall x,y \in \kappa} \bar{x}\langle y \rangle.\bar{i}\langle \kappa \rangle \mid \prod_{\forall x \in \kappa} x(z).\bar{i}\langle \kappa \cup \{z\} \rangle \mid (\nu net)\bar{i}\langle \kappa \cup \{net\} \rangle))$$

In this specification, κ is a set of names representing the knowledge of the intruder, (νnet) allows for the intruder to create fresh data at any time and i is a channel used for the intruder's internal communications. The initial subprocess, $\bar{i}\langle \kappa_{init} \rangle$, outputs the set of names, κ_{init}, representing an instantiation of the intruder's initial knowledge (in general, $\kappa_{init} = fn(P)$, for the analysed process, P).

The above specification allows the intruder to build its knowledge, κ, by repeatedly inputting over names in its knowledge. The inputted name is then passed as part of the new knowledge to the next instance of the intruder. The intruder can also perform output actions. These are either free output actions sending messages over channels already in κ, or bound output actions that create a copy of the name *net* and send it over the internal channel i. This allows the intruder to learn *net* without the need to output it first to external processes. The *learning* behaviour is interpreted as the standard union, $\cup$, over κ.

Assuming the intruder's initial knowledge is set to $\kappa_{init} = \{login, logout\}$, we obtain the following least fixed-point value for $\phi_{\mathcal{A}}$ by applying $\mathcal{A}([ftp]) \{ \} \phi_{\mathcal{A}0}$, where the integer constraint is set here to $k = 1$, signifying a uniform analysis:

$$\phi_{\mathcal{A}} = \begin{bmatrix} \kappa & \mapsto & \{\underline{net_1, pwd_1, request_1}, logout, login_1\} \\ x_1 & \mapsto & \{start, I\_start\} \\ z_1 & \mapsto & \{pwd_1, request_1, logout, login_1, net_1\} \\ u_1 & \mapsto & \{pwd_1, request_1, logout, login_1, net_1\} \\ w_1 & \mapsto & \{pwd_1, request_1, logout, login_1, net_1\} \\ data_1 & \mapsto & \{pwd_1, request_1, logout, login_1, net_1\} \\ res_1 & \mapsto & \{pwd_1, request_1, logout, login_1, net_1\} \end{bmatrix}$$

We have used underlining to indicate the presence of abnormal information in the intruder's knowledge in the final $\phi_{\mathcal{A}}$ environment. The interpretation detects that the intruder is capable of obtaining the *pwd* and *request* names, which is a dangerous result. This result is a false positive since the uniform interpretation does not distinguish between the different copies of the *login* channels. Therefore, it appears to be possible for the intruder to capture any of the other names outputted over different *login* names. Moreover, it is impossible to distinguish between the information each instance of the client and server processes obtains. For example, the *res* input parameter captures not only a *request*, but also the *pwd*, *logout*, *login* and *net* messages. As $\phi_{\mathcal{A}}$ is uniform, it is impossible to distinguish between different sessions.

To refine the above results, we increase the value of k from 1 to 4, i.e. generate a non-uniform analysis. The intermediate values of $k = 2$ and $k = 3$ will still yield the above false positives for $k = 1$ (since the intruder's session interferes with clients' sessions), therefore, we do not include them here. However, when performing the abstract interpretation for $k = 4$, we obtain the following results:

$$\phi_{\mathcal{A}} = \begin{bmatrix} \kappa & \mapsto & \{net_i, login_4, logout \mid i = 1, 2, 3, 4\} \\ x_1 \mapsto \{start\} & x_2 \mapsto \{start\} & x_3 \mapsto \{start\} & x_4 \mapsto \{I\_start\} \\ z_1 \mapsto \{pwd_1\} & z_2 \mapsto \{pwd_2\} & z_3 \mapsto \{pwd_3\} \\ z_4 \mapsto \{net_i, login_4, logout \mid i = 1, 2, 3, 4\} \\ u_1 \mapsto \{logout\} & u_2 \mapsto \{logout\} & u_3 \mapsto \{logout\} & u_4 \mapsto \{\} \\ w_1 \mapsto \{request_1\} & w_2 \mapsto \{request_2\} & w_3 \mapsto \{request_3\} & w_4 \mapsto \{\} \\ data_1 \mapsto \{request_1\} & data_2 \mapsto \{request_2\} & data_3 \mapsto \{request_3\} & data_4 \mapsto \{\} \\ res_1 \mapsto \{request_1\} & res_2 \mapsto \{request_2\} & res_3 \mapsto \{request_3\} \end{bmatrix}$$

The results of the non-uniform analysis ($k = 4$) reveal that the intruder is incapable of obtaining information from sessions involving clients. It also shows that the server, in fact, does not process any request from the intruder, since the latter lacks the required password (in this case, pwd_4). Also, a more precise distribution of information between the three copies of the client and server processes is reflected in the results. For example, it is clear that a client can only receive a *request* message as a final result held by the *res* input parameter. Also, each copy of res_i captures the corresponding copy of $request_i$, for $i = 1, 2, 3$.

Another version of the FTP system may contain a faulty client specified as follows:

$$Client \stackrel{\text{def}}{=} (v\,request)(\overline{login}\langle pwd\rangle.\overline{login}\langle request\rangle.login(res).\overline{login}\langle logout\rangle.\overline{covert}\langle pwd\rangle)$$

The specification of the client reveals, over a *covert* channel, the password it shares with the server to the intruder. Now, if we perform the abstract interpretation for $k = 1$ setting the knowledge of the intruder to $\kappa_{init} = \{covert, login, logout\}$, we obtain the following results:

$$\phi_{\mathcal{A}} = \begin{bmatrix} \kappa & \mapsto \{net_1, \underline{pwd_1}, \underline{request_1}, logout, login_1, covert\} \\ x_1 & \mapsto \{start, I\_start\} \\ z_1 & \mapsto \{pwd_1, request_1, logout, login_1, net_1, covert\} \\ u_1 & \mapsto \{pwd_1, request_1, logout, login_1, net_1, covert\} \\ w_1 & \mapsto \{pwd_1, request_1, logout, login_1, net_1, covert\} \\ data_1 & \mapsto \{pwd_1, request_1, logout, login_1, net_1, covert\} \\ res_1 & \mapsto \{pwd_1, request_1, logout, login_1, net_1, covert\} \end{bmatrix}$$

The intruder obtains the *pwd* and *request* names. Refining the interpretation further by setting $k = 4$ reveals that the intruder, in fact, can only capture the password (when the client reveals it over the *covert* channel):

$$\phi_{\mathcal{A}} = \begin{bmatrix} \kappa & \mapsto \Delta \; where, \; \Delta = \{net_i, login_4, logout, covert, \underline{pwd_j} \mid \\ & \qquad\qquad i = 1, 2, 3, 4 \text{ and } j = 1, \underline{2, 3}\} \\ x_1 \mapsto \{start\} & x_2 \mapsto \{start\}. & x_3 \mapsto \{start\} & x_4 \mapsto \{I\_start\} \\ z_1 \mapsto \{pwd_1\} & z_2 \mapsto \{pwd_2\} & z_3 \mapsto \{pwd_3\} \\ z_4 \mapsto \Delta \\ u_1 \mapsto \{logout\} & u_2 \mapsto \{logout\} & u_3 \mapsto \{logout\} & u_4 \mapsto \{\} \\ w_1 \mapsto \{request_1\} & w_2 \mapsto \{request_2\} & w_3 \mapsto \{request_3\} & w_4 \mapsto \{\} \\ data_1 \mapsto \{request_1\} & data_2 \mapsto \{request_2\} & data_3 \mapsto \{request_3\} & data_4 \mapsto \{\} \\ res_1 \mapsto \{request_1\} & res_2 \mapsto \{request_2\} & res_3 \mapsto \{request_3\} \end{bmatrix}$$

The results of the non-uniform interpretation reveal that the intruder is incapable of compromising the system, even in the case where it compromises the password of each client. This is due to the fact that these passwords are one-time passwords. The name *request* does not appear in any of the values for κ (the intruder's knowledge). Also, the session between the server and the intruder (for z_4) indicates that the intruder is incapable of processing its data by the server, since it cannot pass beyond the login stage.

We next present security properties based on the results of the abstract interpretation of the FTP system. More precisely, we reason about name substitutions and their impact on the secrecy and authenticity of the system.

3.3.1.1   Secrecy

The problem of protecting the secrecy of private information is a long-standing problem that finds its roots in the pre-computing age. Models were devised to prevent unauthorised subjects from accessing classified physical resources (like paperwork) owned by other subjects. The success of these models in the real world encouraged researchers to adopt the same ideas to protect the secrecy of information in the computing world. We adopt in what follows, a form of explicit information flow that monitors the movement of data among the different processes in a system. A low-level process obtaining data created by another high-level process results in an instance of the process leakage threat. The explicit meaning of the word "obtaining" here is strictly defined in terms of the message-passing communications in systems.

In our formalisation of the secrecy property, we assume the presence of predefined multilevel security policies that are determined by the administrators of the systems under analysis. To express these policies, we assume a finite lattice of *secrecy levels*, $L = (S_L, \sqsubseteq, \sqcap, \sqcup, \top, \bot)$, ranged over by $l, l' \in S_L$ with $\bot_L$ being the undefined level and $\top_L$ being the most sensitive level. A security policy then attempts to classify processes according to their sensitivity by assigning to each process its appropriate level. We use the notation $\lceil P \rceil^l$ to express the fact that process P is classified (running) at secrecy level, l. The approach to process classification is flexible but requires that every subprocess within the main specification be assigned some secrecy level. Usually, the bottom level, $\bot_L$, is preserved for the intruder's specification (for example, as specified in the previous section), as this would be the safest assumption to make about the intruder's secrecy property.

In the following, we discuss a variation of the secrecy property, termed *process leakage*. A process leakage occurs whenever a process classified at a low secrecy level explicitly obtains high-level data through one of its input parameters or variables. The meaning of *explicitly obtains* depends on the choice of language. In our language, a process obtains data solely through communications (instantiating input parameters).

In order to be able to analyse the secrecy property, we need to translate process classifications to name classifications. That is to say, given a process, $\lceil P \rceil^l$, running at secrecy level, l, we need to know the levels of its names, $n(P)$. Such translation requires prior knowledge of the level of the intruder running in parallel with the analysed process. As we mentioned in the introduction, the safest level given to such an intruder is $\bot_L$.

For this purpose, we define an environment $\zeta : N \rightarrow L$ that maps names to their secrecy levels. Initially, ζ_0 maps every name to the bottom level. Hence

$$\forall x \in N : \zeta_0(x) = \bot_L$$

We also define the union of ζ environments as follows:

$$\forall x \in N : (\zeta_1 \cup_\zeta \zeta_2)(x) = \zeta_1(x) \sqcup \zeta_2(x)$$

To obtain secrecy levels for the bound names of process P, a special encoding function $\mathcal{Z} : \mathcal{P} \times (N \rightarrow A) \rightarrow (N \rightarrow A)$ is needed to construct the final value for ζ. This function is defined in Fig. 3.2.

The most interesting cases are those for the input actions, name restrictions and process definitions, where the bound name, y, is assigned the level of its process, l. As

$$\mathcal{Z}(\lceil 0 \rceil^l)\,\zeta = \zeta$$
$$\mathcal{Z}(\lceil \text{timer}^{t+1}(x(y).P,Q)\rceil^l)\,\zeta = \mathcal{Z}(P)\,\zeta\,[y \mapsto l]\ \cup_\zeta\ \mathcal{Z}(Q)\,\zeta$$
$$\mathcal{Z}(\lceil \text{timer}^{t+1}(x().P,Q)\rceil^l)\,\zeta = \mathcal{Z}(P)\,\zeta\ \cup_\zeta\ \mathcal{Z}(Q)\,\zeta$$
$$\mathcal{Z}(\lceil \text{timer}^0(x(y).P,Q)\rceil^l)\,\zeta = \mathcal{Z}(Q)\,\zeta$$
$$\mathcal{Z}(\lceil \text{timer}^0(x().P,Q)\rceil^l)\,\zeta = \mathcal{Z}(Q)\,\zeta$$
$$\mathcal{Z}(\lceil \overline{x}\langle y\rangle.P\rceil^l)\,\zeta = \mathcal{Z}(P)\,\zeta$$
$$\mathcal{Z}(\lceil \overline{x}\langle\rangle.P\rceil^l)\,\zeta = \mathcal{Z}(P)\,\zeta$$
$$\mathcal{Z}(\lceil \tau.P\rceil^l)\,\zeta = \mathcal{Z}(P)\,\zeta$$
$$\mathcal{Z}(\lceil P[x=y]Q\rceil^l)\,\zeta = \mathcal{Z}(P)\,\zeta\ \cup_\zeta\ \mathcal{Z}(Q)\,\zeta$$
$$\mathcal{Z}(\lceil P+Q\rceil^l)\,\zeta = \mathcal{Z}(P)\,\zeta\ \cup_\zeta\ \mathcal{Z}(Q)\,\zeta$$
$$\mathcal{Z}(\lceil P \mid Q\rceil^l)\,\zeta = \mathcal{Z}(P)\,\zeta\ \cup_\zeta\ \mathcal{Z}(Q)\,\zeta$$
$$\mathcal{Z}(\lceil (\nu y)P\rceil^l)\,\zeta = \mathcal{Z}(P)\,\zeta\,[y \mapsto l]$$
$$\mathcal{Z}(\lceil !P\rceil^l)\,\zeta = \mathcal{Z}(P)\,\zeta$$
$$\mathcal{Z}(\lceil A(x)\rceil^l)\,\zeta = \mathcal{Z}(P)\,\zeta\,[y \mapsto l]\ \ \text{where,}\ A(y) \stackrel{\text{def}}{=} P$$
$$\mathcal{Z}(\lceil A()\rceil^l)\,\zeta = \mathcal{Z}(P)\,\zeta\ \ \text{where,}\ A() \stackrel{\text{def}}{=} P$$

Fig. 3.2 Rules of the $\mathcal{Z}(\lceil P\rceil^l)\,\zeta$ function

we mentioned earlier, input parameters are treated in the process leakage property
as the means by which processes gain data, whereas restricted names are treated
as local data. Furthermore, the computation of $\mathcal{Z}(\lceil P\rceil^l)\,\zeta$ will terminate since the
process on the right-hand side of each rule is always a subprocess of the process
on the left-hand side. For simplicity, we have chosen a uniform definition for the ζ
function, in the sense that all the copies of bound names of a process are assigned
a single level, which is the level of the root name appearing in the specification. A
more flexible classification would be to assign different levels to different copies, or
in other words, to allow the level of a name to change during runtime.

The following predicate formalises the *process leakage* property. The predicate
takes as input a process, P, analyses it and returns true whenever a high-level name, x,
is *leaked* to some low-level input parameter, y (as a result of some communication).

Property 3.1 (Process leakage) *A name, x, is said to be leaked within a process, P,
if and only if the following holds true:*
$$\phi_{\mathcal{A}} = \mathcal{A}(\lceil P \rceil)\,\rho_0\,\phi_{\mathcal{A}0},\ \zeta = \mathcal{Z}(\lceil P\rceil^l)\,\zeta_0,\ \exists y \in dom(\phi_{\mathcal{A}}),\ \exists x \in \phi_{\mathcal{A}}(y):$$
$$\zeta(y) \sqsubseteq_L \zeta(x) \hspace{4cm} \square$$

where ρ_0 may contain the intruder, I, in the case it is running in parallel with P—
meaning that the environment within which P is running is hostile. $dom(\phi_{\mathcal{A}})$ is the
domain of the environment, $\phi_{\mathcal{A}}$. Informally, the property says that process leakage
occurs whenever some low-level process manages to obtain a name, which was
originally created by some other process with a higher secrecy level.

Going back to the FTP server example, we analysed the system for two cases:
The first case has a correct specification for the client process, which communicates
with the server without interference from the intruder, whereas the second case
has a faulty client specification that leaks its password to the intruder process, I.

To understand what the analysis means for process leakage property, we adopt the following classification of secrecy levels:

$$ftp \stackrel{\text{def}}{=} \lceil (!start(x).(\nu\, login)(if\, x = start\, then\, \lceil (\nu\, pwd)(Server \mid Client)\rceil^{l}$$
$$else\, \lceil (\nu\, pwd)(Server)\rceil^{l} \mid \lceil I \rceil^{\perp_{L}})$$
$$) \mid \overline{start}\langle start\rangle.\overline{start}\langle start\rangle.\overline{start}\langle start\rangle.\overline{start}\langle I\_start\rangle \rceil^{\perp_{L}}$$

$$Server \stackrel{\text{def}}{=} (\nu\, deal)\, login(z).if\, z = pwd\, then$$
$$login(data).($$
$$\overline{deal}\langle data\rangle \mid !deal(w).\overline{login}\langle w\rangle.login(u).if\, u = logout\, then\, 0$$
$$else\, \overline{deal}\langle u\rangle\,)\, else\, 0$$

$$Client \stackrel{\text{def}}{=} (\nu\, request)\, (\overline{login}\langle pwd\rangle.\overline{login}\langle request\rangle.login(res).\overline{login}\langle logout\rangle)$$

with $\perp_{L} \sqsubseteq l$. By applying the encoding function $\mathcal{Z}$, it is possible to obtain the following value for the ζ environment, mapping each name to its secrecy level:

$$\zeta(\kappa) = \zeta(x) = \zeta(net) = \zeta(login) = \zeta(logout) = \zeta(start) = \zeta(I\_start) = \perp_{L}$$
$$\zeta(pwd) = \zeta(deal) = \zeta(request) = l$$

The results of the uniform abstract interpretation (correct client specification) with α_{1} indicate that the intruder is capable of capturing the *pwd* and *request* names, both of which have a higher secrecy level than κ. Refining these results further, we find that the non-uniform abstract interpretation, with α_{4}, reveals a correct distribution of names to input parameters in the final value for $\phi_{\mathcal{A}}$. In particular, the intruder's knowledge, κ, could not obtain any names with higher secrecy levels and remained limited to names *net*, *login* and *logout*. In general, $\forall x \in dom(\phi_{\mathcal{A}}), y \in \varphi_{\mathcal{A}}(\phi_{\mathcal{A}}, x) :$ $\zeta(y) \sqsubseteq_{L} \zeta(x)$.

However, examining the results of the second non-uniform abstract interpretation (faulty client specification) with α_{4}, the analysis captures an instance of a process leakage as defined by Property 3.1. This secrecy breach occurs with the value of κ, which captures the secret password, *pwd*. Since $\zeta(pwd) = l$ and $\zeta(\kappa) = \perp_{L}$ according to the above name classification, we have that $\zeta(\kappa) \sqsubseteq \zeta(pwd)$. It is clear that this breach happened as a result of the client sending its password over the *covert* channel, which is a free name recognised by the intruder.

3.3.1.2 Authenticity

The authenticity property is generally defined as the assurance about the identity of the origin of a particular entity (agent, datum, message, etc.). This identification may entail some *trust* in the sense that identifying a trusted entity could initiate different behaviour than in the event of identifying a less trusted entity. For example, when performing online booking for air flight tickets, we would like to make sure that the booking site, to which the credit card details are submitted, has a valid digital certificate signed by a trusted third party (or a Certification Authority). This will increase the confidence about sending credit card details to the site. However, if no valid certificates were produced, then the site will be less trustworthy in submitting sensitive information. The model of authentication we adopt here is based on the assumption that systems are classified as running at different trust levels. Breaches

of authenticity will occur whenever highly trusted processes obtain data originating
at processes with lower levels of trust.

The treatment of the process leakage property in the previous section was made
possible by using the notion of secrecy levels that distinguished the secrecy requirements of each process according to a well-defined security policy. In dealing with the
authenticity property, the main notion of interest is that of process *trust level*. Trust
levels, $a, a' \in S_A$, are ordered by a web of trust such that $A = (S_A, \sqsubseteq, \sqcap, \sqcup, \top, \bot)$
is a finite lattice with the bottom element, $\bot_A$, being the undefined level and the
top element, $\top_A$, being the most trusted level. Examples of undefined trust levels
include the levels of machines connected to the Internet, which belong to unknown
entities. A well-trusted level, on the other hand, could be a Certification Authority
(CA) in a Public Key Infrastructure (PKI). The work of [2] represents one example
in which mechanisms for implementing trust levels, called *addresses*, are provided
as a primitive in an extension of the π-calculus language.

As a result of the presence of malicious attackers, a threat of process authenticity may occur in situations where a process obtains data that originated at a trust
level lower than the level of the process itself. Intuitively, such threats are directly
comparable to process leaks. However, with the process leakage property, it is the
high-level data that are compromised by a secrecy breach (being obtained by a low-
level process), whereas with the process authenticity property, it is the high-trust
process that is compromised by an authenticity breach (obtaining the low-trust data).
We emphasise here the direction of concern, which is different in each case.

In what follows, we write $\lfloor P \rfloor^a$ to express the fact that P is running at trust level a.
The requirement then is to translate process trust levels to name trust levels assuming
that all subprocesses Q of P have the form, $\lfloor Q \rfloor^{a'}$, for some level a'. Furthermore, we
translate process trust levels to name trust levels using the environment, $\theta : N \rightarrow A$,
which maps names to their levels. Initially, θ_0 maps every name to the bottom element,
$a_I = \bot_A$, also assumed to be the level of I:

$$\forall x \in N : \theta_0(x) = \bot_A$$

Additionally, we define the union of θ-environments, $\cup_\theta$, as follows:

$$(\theta_1 \cup_\theta \theta_2)(x) = \theta_1(x) \sqcup \theta_2(x)$$

This is similar to the union of ζ-environments defined in the previous section on
secrecy. The use of the least upper bound operator $\sqcup$ is useful for adding any values
of a name from θ_1 and θ_2 that are equal to $\bot_A$.

Defining a special encoding function, $\mathcal{U} : \mathcal{P} \times (N \rightarrow A) \rightarrow (N \rightarrow A)$, as in
Fig. 3.3, is necessary to construct θ from the specification of $\lfloor P \rfloor^a$.

The rules of the $\mathcal{U}(\lfloor P \rfloor^a) \, \theta$ function are directly comparable to their secrecy
counterpart given by the $\mathcal{Z}(\lceil P \rceil^l) \, \zeta$ function (in fact, the two functions only differ
in whether they deal with secrecy or trust levels). The treatment of bound names is
similar. These are assigned the trust levels of their process. In the case of restricted
names, this reflects their locality, whereas in the case of input parameters, this reflects

$$\mathcal{U}(\lfloor 0 \rfloor^a)\ \theta = \theta$$
$$\mathcal{U}(\lfloor \mathtt{timer}^{t+1}(x(y).P,Q) \rfloor^a)\ \theta = \mathcal{U}(P)\ \theta[y \mapsto l]\ \cup_\theta\ \mathcal{U}(Q)\ \theta$$
$$\mathcal{U}(\lfloor \mathtt{timer}^{t+1}(x().P,Q) \rfloor^a)\ \theta = \mathcal{U}(P)\ \theta\ \cup_\theta\ \mathcal{U}(Q)\ \theta$$
$$\mathcal{U}(\lfloor \mathtt{timer}^{0}(x(y).P,Q) \rfloor^a)\ \theta = \mathcal{U}(Q)\ \theta$$
$$\mathcal{U}(\lfloor \mathtt{timer}^{0}(x().P,Q) \rfloor^a)\ \theta = \mathcal{U}(Q)\ \theta$$
$$\mathcal{U}(\lfloor \overline{x}\langle y \rangle.P \rfloor^a)\ \theta = \mathcal{U}(P)\ \theta$$
$$\mathcal{U}(\lfloor \overline{x}\langle \rangle.P \rfloor^a)\ \theta = \mathcal{U}(P)\ \theta$$
$$\mathcal{U}(\lfloor \tau.P \rfloor^a)\ \theta = \mathcal{U}(P)\ \theta$$
$$\mathcal{U}(\lfloor P[x=y]Q \rfloor^a)\ \theta = \mathcal{U}(P)\ \theta\ \cup_\theta\ \mathcal{U}(Q)\ \theta$$
$$\mathcal{U}(\lfloor P+Q \rfloor^a)\ \theta = \mathcal{U}(P)\ \theta\ \cup_\theta\ \mathcal{U}(Q)\ \theta$$
$$\mathcal{U}(\lfloor P\mid Q \rfloor^a)\ \theta = \mathcal{U}(P)\ \theta\ \cup_\theta\ \mathcal{U}(Q)\ \theta$$
$$\mathcal{U}(\lfloor (\nu y)P \rfloor^a)\ \theta = \mathcal{U}(P)\ \theta[y \mapsto l]$$
$$\mathcal{U}(\lfloor !P \rfloor^a)\ \theta = \mathcal{U}(P)\ \theta$$
$$\mathcal{U}(\lfloor A(x) \rfloor^a)\ \theta = \mathcal{U}(P)\ \theta[y \mapsto l]\quad \text{where, } A(y) \stackrel{\mathrm{def}}{=} P$$
$$\mathcal{U}(\lfloor A() \rfloor^a)\ \theta = \mathcal{U}(P)\ \theta\quad \text{where, } A() \stackrel{\mathrm{def}}{=} P$$

Fig. 3.3 Rules of the $\mathcal{U}(\lfloor P \rfloor^a)\ \theta$ function

the fact that they are used as means by which the residual process acquires further information (names). All the other cases do not affect the value of θ.

The following property formalises the process authenticity breach.

Property 3.2 (Process authenticity breach) *The authenticity of a name, y, is said to be breached within a process, P, if and only if the following holds true:*
$$\phi_{\mathcal{A}} = \mathcal{A}(\![P]\!)\ \rho_0\ \phi_{\mathcal{A}0},\ \theta = \mathcal{U}(\lfloor P \rfloor^a)\ \theta_0,\ \exists y \in dom(\phi_{\mathcal{A}}),\ \exists x \in \phi_{\mathcal{A}}(y):\ \theta(y) \sqsupseteq \theta(x)\ \square$$

where ρ_0 may contain the intruder, I. When compared to the definition of the process leakage property (Property 3.1), the direction of the ordering relation clearly demonstrates the difference in concern. Intuitively, a process authenticity breach occurs whenever a tag, whose name value has a low trust level, instantiates another name with a higher trust level. Hence, we are concerned with highly trusted processes obtaining data not at the same level of trust. This could be a result of that data originating from malicious sources.

Considering our FTP example, we adopt the following classification for trust levels:

$$ftp \stackrel{\mathrm{def}}{=} \lfloor (!start(x).(\nu login)(if\ x = start\ then\ \lfloor (\nu pwd)(Server \mid Client) \rfloor^a$$
$$else\ \lfloor (\nu pwd)(Server) \rfloor^a \mid \lfloor I \rfloor^{\perp_A}$$
$$)$$
$$)\mid \overline{start}\langle start \rangle.\overline{start}\langle start \rangle.\overline{start}\langle start \rangle.\overline{start}\langle I\_start \rangle \rfloor^{\perp_A}$$

$$Server \stackrel{\mathrm{def}}{=} (\nu deal)\ \lfloor login(z).\lfloor if\ z = pwd\ then$$
$$login(data).($$
$$\overline{deal}\langle data \rangle \mid !deal(w).\overline{login}\langle w \rangle.login(u).if\ u = logout\ then\ 0$$
$$else\overline{deal}\langle u \rangle\ )\ else0 \rfloor^a \rfloor^{\perp_A}$$

$$Client \stackrel{\mathrm{def}}{=} (\nu request)\ (\overline{login}\langle pwd \rangle.\overline{login}\langle request \rangle.login(res).\overline{login}\langle logout \rangle)$$

with $\perp_A \sqsubseteq a$. Applying the encoding function $\mathcal{U}$ to each of the above (sub)processes, it is possible to construct a θ environment mapping names to their trust levels, as follows:

$\theta(\kappa) = \theta(x) = \theta(z) = \theta(net) = \theta(login) = \theta(logout) = \theta(start) = \theta(I\_start) = \perp_A$
$\theta(pwd) = \theta(deal) = \theta(request) = a$

In both analyses (with correct and faulty client specifications) with α_4, we find that input parameters captured the appropriate names. In general, $\forall x \in dom(\phi_\mathcal{A}), y \in \varphi_\mathcal{A}(\phi_\mathcal{A}, x) : \theta(x) \sqsubseteq_A \theta(y)$. Hence, no instance of the process authenticity breach occurs (Property 3). The case of the faulty client is interesting, since the intruder, I, failed in passing its low-level data to the clients, even though it clearly breaches the secrecy of their passwords as indicated above. This is due to the fact that the passwords are per login session, i.e. each time a copy of the system is spawned, the passwords and the login channel are renamed. Hence, the intruder cannot use a password it obtains from the client in its own session with the FTP server. Attempting to do that means that it will fail as a result of the conditional process that the server runs and that maintains a correct distribution of password-login session names. Hence, the authenticity property is preserved despite a failure in the secrecy property of the protocol.

3.3.2 Simple Example 2: A Distance-Bounding Protocol

We consider next a second simple example to demonstrate how the abstract interpretation can be used to understand better the behaviour of protocols, again from a security point of view. The protocol we consider is a simplified version of the RFID distance-bounding protocol defined in [3]. Distance-bounding protocols have the requirement that a message arrives at its destination in a timely fashion. For example, consider the following scenario:

$$@T_1 : Alice \xrightarrow{m_1} @T_2 : Carol$$

$$@T_3 : Carol \xrightarrow{m_2} @T_4 : Bob$$

in which Alice sends a message m_1 to Bob at time T_1. The message is intercepted by Carol at time T_2, altered to m_2 and then forwarded to Bob at time T_3. Finally, Bob receives m_2 at time T_4. If Bob is time-sensitive, he would have two parameters: the first is T_{exp}, the time at which he expects to receive the message, and the second is Diff, the maximum difference he is willing to tolerate between the actual time of receipt and T_{exp}. Therefore, in the above scenario, one would expect Bob to time out and reject m_2 as inauthentic if $|T_4 - T_{exp}| > Diff$. Otherwise, if Carol succeeds in modifying m_1 in a timely fashion, i.e. such that $|T_4 - T_{exp}| \leq Diff$, then Bob is likely to accept the message as authentic (if all other non-time-related criteria are satisfied).

The one-way authentication protocol of [3] consists of the following steps between a Verifier, Vr, and a Prover, Pr, starting at time, T_0:

$$Vr \stackrel{\text{def}}{=} (\nu N_{Vr})(\nu x) \ (\overline{c}\langle N_{Vr}\rangle.\texttt{timer}^{T1}(x(d1).\mathbf{0}, \overline{c}\langle C1\rangle.$$
$$\texttt{timer}^{\delta}(c(r1).\texttt{timer}^{T2}(x(d2).\mathbf{0}, \overline{c}\langle C2\rangle.$$
$$\texttt{timer}^{\delta}(c(r2).\texttt{timer}^{T3}(x(d3).\mathbf{0}, \overline{c}\langle C3\rangle.$$
$$\texttt{timer}^{\delta}(c(r3).\mathbf{0}, \mathbf{0})\,), \mathbf{0})\,), \mathbf{0})\,)\,)$$

$$Pr \stackrel{\text{def}}{=} \texttt{timer}^{\infty}(c(n).\texttt{timer}^{\infty}(c(u1).\overline{c}\langle R1\rangle.\texttt{timer}^{\infty}(c(u2).\overline{c}\langle R2\rangle.$$
$$\texttt{timer}^{\infty}(c(u3).\overline{c}\langle R3\rangle.\mathbf{0}, \mathbf{0}), \mathbf{0}), \mathbf{0}), \mathbf{0})$$

$$I \stackrel{\text{def}}{=} (\nu N_I) \ (\texttt{timer}^{\infty}(c(n').\overline{c}\langle N_I\rangle.$$
$$\texttt{timer}^{\infty}(c(u1').\overline{c}\langle C1'\rangle.\texttt{timer}^{\infty}(c(r1').\overline{c}\langle R1'\rangle.$$
$$\texttt{timer}^{\infty}(c(u2').\overline{c}\langle C2'\rangle.\texttt{timer}^{\infty}(c(r2').\overline{c}\langle R2'\rangle.$$
$$\texttt{timer}^{\infty}(c(u3').\overline{c}\langle C3'\rangle.\texttt{timer}^{\infty}(c(r3').\overline{c}\langle R3'\rangle.\mathbf{0}, \mathbf{0}), \mathbf{0}), \mathbf{0}), \mathbf{0}), \mathbf{0}))$$

$$Prot \stackrel{\text{def}}{=} I \mid (Vr \mid Pr)$$

Fig. 3.4 The definition of the RFID protocol of [3]

$$@T_0 : \qquad\qquad Vr \to Pr : N_{Vr}$$
$$for(i = 1; i \leq n; inc(i)) \{$$
$$@T_i : \qquad\quad Vr \to Pr : C_i$$
$$@(T_i + \delta) : \quad Pr \to Vr : R_i^{C_i} \ \}$$

where $n > 0$, $T_i, \delta \in \mathbb{N}$ are natural numbers such that T_i is a point in time and δ is a very short time gap (ideally $T_i + \delta < T_{i+1}$). Also, $inc : \mathbb{N} \to \mathbb{N}$ is the increment function, N_{Vr} is a fresh nonce and C_i, $R_i^{C_i}$ are challenge values and their corresponding responses. For the sake of brevity, we refer the reader for a full description of the protocol to [3, Sect. 3.1]. Here, we give in Fig. 3.4 a non-cryptographic specification of the protocol for the specific case of $n = 3$.

The specification allows Vr to send a fresh nonce N_{Vr} to Pr. Vr then uses the internal channel x to simulate time waitings of T_1, T_2 and T_3 since no inputs can be performed over x and these will time out. However, their continuations will output challenges Ci to Pr. Pr itself waits on these challenges and then replies with the expected responses Ri. During this protocol, the intruder I is capable of interfering with all communications over c, since it knows the name of this channel. The protocol itself is defined as the parallel composition of the three processes.

Applying the abstract interpretation, $\mathcal{A}(\llbracket Prot \rrbracket) \{\!|\!\} \perp_{D_\perp^\sharp}$, with $k = 1$, we obtain the following substitutions for $i = 1 \ldots n$:
$$Ci' \in \phi_{\mathcal{A}}(ui) \text{ and } Ri' \in \phi_{\mathcal{A}}(ri)$$
In order to understand the security implications of the analysis, we first refer to the usual finite lattice of security levels, $(\mathcal{S}, \sqsubseteq_S, \sqcap_S, \sqcup_S, \top_S, \perp_S)$. Based on the definition of this lattice, we define $\zeta : \mathcal{N} \to \mathcal{S}$ as a mapping from names to their security levels. We can now define an interesting property, *name integrity*, as follows.

Property 3.3 (Name integrity) *We say that a name, x, has the integrity property with respect to a $\phi_{\mathcal{A}}$ environment if* $\forall n \in value\_of(\phi_{\mathcal{A}}(x)) : \zeta(x) \sqsubseteq \zeta(n)$ $\qquad\square$

The predicate $integrity(x, \phi_{\mathcal{A}})$ indicates that x upholds the above property with respect to $\phi_{\mathcal{A}}$. A Man-in-the-Middle (MitM) attack is defined as an attack in which the intruder is capable of breaching the integrity of the names of two processes.

$$I \stackrel{\text{def}}{=} (\nu i) \, (! \, \texttt{timer}^{t_1} \, (i(\kappa).(\prod_{\forall x, y \in \kappa} \bar{x}^{t_2} \langle y \rangle. \bar{i}^{t_3} \langle \kappa \rangle \mid$$

$$\prod_{\forall x \in \kappa} \texttt{timer}^{t_4} \, (x(z).\bar{i}^{t_5} \langle \kappa \cup \{z\} \rangle, \mathbf{0}) \mid (\nu net) \bar{i}^{t_6} \langle \kappa \cup \{net\} \rangle)), \mathbf{0})$$

Fig. 3.5 Specification of the Dolev–Yao Intruder

Property 3.4 (Man-in-the-Middle Attack) *An intruder, I, succeeds in launching an MitM attack on two processes, P and Q, if the result of the abstract interpretation, $\mathcal{A}[\![P \mid Q \mid I]\!] \{\![\}\!] \perp_{D^{\sharp}_{\perp}} = \phi_{\mathcal{A}}$, proves that*
$\exists x \in bn(P), y \in bn(Q) : \neg(integrity(x, \phi_{\mathcal{A}}) \vee integrity(y, \phi_{\mathcal{A}}))$ $\qquad\qquad$ □

Now, assuming that the intruder's challenges and responses have lower security levels than the prover's and verifier's input parameters, i.e. $\zeta(Ci') \sqsubseteq \zeta(ui)$ and $\zeta(Ri') \sqsubseteq \zeta(ri)$, then it can be seen that I achieves the MitM property above with respect to ui and ri. This is due mainly to the promptness with which I sends its challenges and responses to both the prover and the verifier processes.

One of the benefits of modelling the intruder as any other process in the specification of the system, rather than, for example, hardcoding its behaviour directly into the semantics of the language, is that it is possible to capture different classes of intruders, ranging from the most passive intruder up to the most general one as envisioned by the Dolev–Yao model of the most powerful intruder [1, 4].

In a closed environment where there are no intruders, one may assume I to be the most passive intruder and model it as the null process $\mathbf{0}$. However, the more interesting extreme is that of the Dolev–Yao intruder. This intruder could be specified as in Fig. 3.5, where $\prod$ denotes the parallel composition of multiple processes.

In this specification, κ denotes a set of names representing the knowledge of the intruder, (νnet) allows for the intruder to create fresh data at any time and i is a channel used for the intruder's internal communications. The initial subprocess, $\bar{i} \langle \kappa_{init} \rangle$, can be used in parallel with I (as in $\bar{i} \langle \kappa_{init} \rangle \mid I$) in order to initialise the intruder's first knowledge, which is the set of names, κ_{init} (in general, $\kappa_{init} = fn(P)$, for the analysed process, P). The specification allows the intruder to build its knowledge, κ, by repeatedly inputting over names in its knowledge. The inputted name is then passed as part of the new knowledge to the next instance of the intruder. The intruder can also perform *timed output actions*, which have the general form:

$$\bar{c}^t \langle m \rangle.P$$

and which are the syntactic sugar for

$$\nu x(\texttt{timer}^t (x(y).\mathbf{0}, \bar{c} \langle m \rangle.P)$$

This means that the output action is delayed by t time before it is available. Some of these outputs are either free output actions sending messages over channels already in κ, or bound output actions that create a copy of the name *net* and send it over the internal channel i. This allows the intruder to learn *net* without the need to output it first to external processes. The *net* names represent information that the intruder is capable of generating internally. The *learning* behaviour is interpreted as the standard union, $\cup$, over κ. The values of the timings t_i for $i = 1 \ldots 6$ will determine the success or failure of the intruder in achieving a precise timing MitM attack on any two processes.

The definition we provided for the Dolev–Yao intruder does not make specific any of the time delays in the input or output actions. However, one extreme version of the Dolev–Yao intruder is *the Fastest Dolev–Yao Intruder*, whose specification is the same as the one presented in Fig. 3.5, except that it has

$$t_2 = t_3 = t_5 = t_6 = 0$$

This means that the intruder is prompt in performing any of its output actions including the ones with external processes over names included in its knowledge. On the other hand, the intruder never times out its input actions:

$$t_1 = t_4 = \infty$$

The dual of the fastest intruder, i.e. the slowest intruder, is effectively equivalent to the passive intruder, **0**. This is because it cannot wait more than 0 time on its inputs, and it will only perform its outputs after ∞ time. Therefore, we do not consider it to be useful in our analysis here.

Slower intruders can also succeed in mounting an MitM attack if they manage to achieve

$$t_4 + t_5 + t_2 + t_1 < \delta$$

in the case where these intruders are attempting to passively redirect the prover's messages to the verifier, or

$$t_6 + t_1 < \delta$$

in case they are attempting to corrupt the verifier with the *net* names.

References

1. D. Dolev, A. Yao, in *Proceedings of the 22nd Annual Symposium on Foundations of Computer Science* (1981), pp. 350–357
2. C. Bodei, P. Dagano, Theoret. Comput. Sci. **283**(2), 271 (2002)
3. G.P. Hancke, M.G. Kuhn, in *Proceedings of the First International Conference on Security and Privacy for Emerging Areas in Communications Networks* (ACM Press, Athens, Greece, 2005), pp. 67–73
4. I. Cervesato, in *Proceedings of the 16th Annual Symposium on Logic in Computer Science*, ed. by J. Halpern (IEEE Computer Society Press, Boston, MA, U.S.A., 2001), pp. 246–265

Chapter 4
First Case Study: The MQTT Protocol

4.1 Introduction

The MQ Telemetry Transport (MQTT) protocol (version 3.1) [1] is described as a lightweight broker-based publish/subscribe messaging protocol that was designed to allow devices with small processing power and storage, such as those which the IoT is composed of, to communicate over low-bandwidth and unreliable networks. MQTT is nowadays used in many business scenarios, for example, the Facebook Messenger [2].

The publish/subscribe message pattern [3], on which MQTT is based, provides for one-to-many message distribution with three varieties of delivery semantics, based on the level of Quality of Service (QoS) expected. The protocol additionally defines the message structure needed in communications between *client*, i.e. end-devices responsible for generating data from their domain (the data source) and *servers*, which are the system components responsible for collating source data from clients/end-devices and distributing these data to interested *subscribers*.

In the "at most once" case, messages are delivered with the best effort of the underlying communication infrastructure, which is usually IP-based, therefore, there is no guarantee that the message will arrive. This protocol, termed the $QoS = 0$ *protocol*, is represented by the following flow of messages and actions:

$$Client \rightarrow Server : \textbf{Publish}$$
$$Server\ Action : Publish\ message\ to\ subscribers$$

In the second case of "at least once" semantics, additional mechanisms are incorporated to allow for message duplication, and despite the guarantee of delivering the message, there is no guarantee that duplicates will be suppressed. This case is represented by the following flow of messages and actions:

© The Author(s), under exclusive license to Springer Nature Switzerland AG 2022
B. Aziz, *Formal Analysis by Abstract Interpretation*, SpringerBriefs in Applied Sciences and Technology, https://doi.org/10.1007/978-3-030-91153-9_4

$$Client \rightarrow Server : \textbf{Publish}$$
$$Client\ Action : Store\ Message$$
$$Server\ Actions : Store\ Message,$$
$$Publish\ message\ to\ subscribers,$$
$$Delete\ Message$$
$$Server \rightarrow Client : \textbf{Puback}$$
$$Client\ Action : Discard\ Message$$

The second message **Puback** represents an acknowledgement of the receipt of the first message, and if **Puback** is lost, then the first message is retransmitted by the client (hence the reason why the message is stored at the client). Once the protocol completes, the client discards the message. This protocol is also known as the $QoS = 1$ *protocol*.

Finally, for the last case of "exactly once" delivery semantics, also known as the $QoS = 2$ *protocol*, the published message is guaranteed to arrive only once at the subscribers. This is represented by the following message/action flow:

$$Client \rightarrow Server : \textbf{Publish}$$
$$Client\ Action : Store\ Message$$
$$Server\ Actions : Store\ Message\ OR$$
$$Store\ Message\ ID,$$
$$Publish\ message\ to\ subscribers$$
$$Server \rightarrow Client : \textbf{Pubrec}$$
$$Client \rightarrow Server : \textbf{Pubrel}$$
$$Server\ Actions : Publish\ message\ to\ subscribers,$$
$$Delete\ Message\ OR$$
$$: Delete\ Message\ ID$$
$$Server \rightarrow Client : \textbf{Pubcomp}$$
$$Client\ Action : Discard\ Message$$

Pubrec and **Pubcomp** represent acknowledgement messages from the server, whereas **Pubrel** is an acknowledgement message from the client. The loss of **Pubrec** causes the client to recommence the protocol from its beginning, whereas the loss of **Pubcomp** causes the client to retransmit only the second part of the protocol, which starts at the **Pubrel** message. This additional machinery ensures a single delivery of the published message to the subscribers.

QoS Level 0 Protocol:

$Client(Publish) \mid Server()$, where:

$Client(z) \overset{\text{def}}{=} \overline{c}\langle z \rangle$

$Server() \overset{\text{def}}{=} c(x).\overline{pub}\langle x \rangle$

QoS Level 1 Protocol:

$Client(Publish) \mid Server()$, where:

$Client(z) \overset{\text{def}}{=} \overline{c}\langle z \rangle.\texttt{timer}^t(c'(y), Client(Publish_{DUP}))$

$Server() \overset{\text{def}}{=} !c(x).\overline{pub}\langle x \rangle.\overline{c'}\langle Puback \rangle$

QoS Level 2 Protocol:

$Client(Publish) \mid Server()$, where:

$Client(z) \overset{\text{def}}{=} \overline{c}\langle z \rangle.\texttt{timer}^t(c(y).ClientCont(y), Client(Publish_{DUP}))$

$ClientCont(u) \overset{\text{def}}{=} \overline{c'}\langle Pubrel_u \rangle.\texttt{timer}^{t'}(c'(w), ClientCont(u))$

$Server() \overset{\text{def}}{=} !c(l).(ServerLate(l) + ServerEarly(l))$

$ServerLate(x) \overset{\text{def}}{=} (\overline{c}\langle Pubrec_x \rangle.c'(v).\overline{pub}\langle x \rangle.\overline{c'}\langle Pubcomp_v \rangle.$
$$!(c'(v').\overline{c'}\langle Pubcomp_{v'} \rangle))$$

$ServerEarly(x) \overset{\text{def}}{=} (\overline{pub}\langle x \rangle.\overline{c}\langle Pubrec_x \rangle.c'(q).\overline{c'}\langle Pubcomp_q \rangle.$
$$!(c'(q').\overline{c'}\langle Pubcomp_{q'} \rangle))$$

Fig. 4.1 A formal model of MQTTv3.1 considering the three levels of QoS

4.2 A Model of the MQTT Protocol

We define a model of the MQTT protocol [1] in our formal language, as shown in Fig. 4.1, which captures the client/server protocol messages. Although the protocol also describes messages between the server and the subscribers, we only focus on one aspect of these, which is the initial publish message from the server to the subscribers. The model expresses three protocols, one for each of the three levels of the quality of service specified in [1].

4.2.1 The Subscribers

Our model of the subscribers is minimal, since we only care about the first step in their behaviour, which is listening to the published messages announced by the server:

$$Subscriber() \overset{\text{def}}{=} !pub(x')$$

This definition does not care about what happens to the message after it has been read by the subscriber on the channel *pub*. The main reason for including the replication, !, is to allow for the possibility of accepting multiple messages from the server. This will allow us later in the analysis to validate the different delivery semantics associated with the MQTT protocol. The definition can capture the number of times the subscriber will read a message within a single run of the protocol since each instance of x' spawned under the replication is renamed with the labelling system x'_1, x'_2, etc. (following [4]). The definition also assumes that the subscriber can wait ad infinitum for a message to be published by the server, and if no such message is published, it will do nothing. This is not realistic, but sufficient for our analysis purposes in this chapter.

4.2.2 The Passive Attacker

The initial model of the attacker that we consider here is one who has a primitive (passive) role; that is, one that offers the possibility of consuming the exchanged messages in the protocol only, particularly those messages between the clients and servers. Therefore, its definition is to listen continuously on the channels c and c' over which the client and the server communicate:

$$Attacker() \overset{\text{def}}{=} \ !(c(y') + c'(u'))$$

Similar to the case of the subscribers, the attacker is not in a rush to obtain an input from the protocol, therefore, it can wait ad infinitum for a message to be received on its channels c or c' (hence we do not use a timer, or instead it is possible to use a timer with $t = \infty$). It is also possible to define an attacker model with finite input capabilities as follows:

$$Attacker() \overset{\text{def}}{=} \ (c(y'_1) + c'(u'_1)) \mid \ldots \mid (c(y'_n) + c'(u'_n))$$

where the operator ! is replaced by a finite number n of the input choices all composed in parallel. In this finite model, the attacker has a linear behaviour in which it is capable of consuming up to only n messages.

In real terms, the case of a passive attacker represents a lossy network acting as a communication medium for MQTT clients and servers. At this stage, we assume that the attacker is not interested in disrupting the server–subscriber part of the protocol communications. However, we discuss this point informally later in Sect. 4.5.

4.3 Analysis of the Protocol

We now define formally the three message-delivery semantics associated with the MQTT protocol, *at most once*, *at least once* and *exactly once* delivery, and we discuss the results of analysing the protocol in light of these three semantics.

4.3.1 QoS = 0 Protocol

The model of QoS $= 0$ protocol is straightforward. The client process is called from the top-level protocol with the *Publish* message. This process is then run in parallel with the server process, which upon receiving the *Publish* message publishes it on the *pub* channel where interested subscribers are listening. For simplicity, we assume that the message is published as is. However, a more refined (but not of interest to us) server process would be expected to extract the relevant payload from *Publish* before publishing the actual data. We formalise the semantics of the protocol for QoS $= 0$ by the following theorem.

Theorem 4.1 (Delivery Semantics For *QoS = 0*) *The MQTT protocol for the case of QoS $= 0$ has a delivery semantics of the publish message to the subscribers of "at most once".*

Proof Given the definition of the subscribers' process in the previous section, a run of this protocol would be equivalent to the following in the absence of any attackers: (*Client*(*Publish*) | *Server*() | *Subscriber*()). Analysing the process renders the following value of ϕ:

$$\phi = \{x \mapsto \{Publish\}, x_1' \mapsto \{Publish\}\}$$

From this, we can see that the message arrives at the subscriber. However, if we re-run the analysis with the attacker process activated, (*Client*(*Publish*) | *Server*() | *Subscriber*() | *Attacker*()), we obtain the following outcome:

$$\phi_{atk} = \{y_1' \mapsto \{Publish\}\}$$

This case shows a run of the protocol, which leads to only y' being instantiated with *Publish*. There is no instantiation of the x or x' variables.

From these results, it is easy to see that there are two possible outcomes. The first value of ϕ represents a normal run where $x' \mapsto \{Publish\}$, whereas in the second value of ϕ_{atk}, we have that $x' \mapsto \{\}$ by the definition of the default state ϕ_0. Hence, it is straightforward to see that the protocol *may* deliver the published message to the subscribers, and therefore, it correctly exhibits the "at most once" delivery semantics. $\square$

For all values of $k > 1$, we do not gain more information about the protocol since the definition of the protocol in Fig. 4.1 does not contain replication, which prevents it from being able to interact with an attacker that generates more messages (i.e. a spammy rather than a lossy attacker).

4.3.2 QoS = 1 Protocol

The QoS = 1 protocol has a semantics of "at least once" delivery. We model this in Fig. 4.1 as a client process, which starts by sending a *Publish* message to the server. The server is capable of inputting this message, publishing it to the subscribers and then replying back to the client with the *Puback* message. Again, for simplicity, we abstract away from the structure of both *Publish* and *Puback*, and point out here that a more refined treatment of these messages (i.e. extracting their payload) does not affect our analysis in the chapter.

The next part is the main difference from the QoS = 0 case above. The client will wait for a finite amount of time, t, on its input channel c' for the *Puback* message from the server. If this message delays (as a result of some communication failure), the client will re-call its process with a new *Publish$_{DUP}$* message. The difference between *Publish$_{DUP}$* and *Publish* is that the DUP bit is set in the former to indicate that it is a duplication of the latter. The server on its side is capable of receiving this new publish message since its behaviour is replicated, which means that it can restart its process any number of times required by the context.

The two channels, c and c', distinguish between the two parts of the protocol (i.e. the *Publish* and *Puback* parts). This is not necessary in practice, however, it renders our model much simpler by avoiding unnecessary interferences between these two parts. In practice, there would be some message validation mechanisms to prevent such interferences from occurring. We formalise the delivery semantics for this case in terms of the following property.

Property 4.1 *(Delivery Semantics For QoS = 1) The MQTT protocol for the case of QoS = 1 has a delivery semantics of the publish message to the subscribers of "at least once".*   □

In order to understand whether this property holds or not (i.e. whether we can promote it to a theorem or not), we need to analyse both cases when the attacker is present and not. We start first by analysing the protocol under the no-attacker conditions. In this case, we find the following subset value for ϕ and $k = 1$:

$$\phi = \{x_1 \mapsto \{Publish\}, y \mapsto \{Puback\}, x_1' \mapsto \{Publish\}\}$$

This implies normal behaviour, where the published message eventually arrives at the subscriber. Therefore, we can state that for the no-attacker conditions, the protocols hold well for the case of QoS = 1. Notably here, the case where $k > 1$ will not add any more information since the server is prompt in replying to the client and so no further calls of the client process definition are made.

Next, we run the analysis with the attacker activated, which produces the following subset value of ϕ_{atk}, where $k = 4$ to initiate multiple runs of the protocol:

$$\phi_{atk} = \{x_1 \mapsto \{\}, u_1' \mapsto \{\}, x_1' \mapsto \{\}, y_1' \mapsto \{Publish\}$$
$$x_2 \mapsto \{Publish\}, u_2' \mapsto \{Puback\}, x_2' \mapsto \{Publish\},$$
$$x_3 \mapsto \{Publish_{DUP}\}, u_3' \mapsto \{Puback\}, x_3' \mapsto \{Publish_{DUP}\},$$
$$x_4 \mapsto \{Publish_{DUP}\}, u_4' \mapsto \{Puback\}, x_4' \mapsto \{Publish_{DUP}\}\}$$

Let's consider first the results for the second, third and fourth copies of the captured variables above. We can see from these that the attacker interferes with the protocol by consuming the *Puback* message. In the next two subsets, therefore, the client issues a duplicate *Publish$_{DUP}$*. In both of these subsets, the attacker continues to consume the acknowledgement message and the client will continue to restart the protocol. Examining these results, we can easily see that the subscribers' input x' has more than one instantiation of the message *Publish*, including when the DUP bit is set. This indicates that the message may arrive more than once at the subscriber. This result is the same for the case of analysis carried out in [5], where only $k = 3$ was considered.

However, let's now examine the first line of results. We notice from this line that the server is not able to receive the message *Publish*. Furthermore, we see that the attacker actually consumes this message instead from the fact that the first copy of y' is instantiated to *Publish*. Analysing this result, we can see that neither the server nor the subscriber will be able to obtain the published message under the circumstances where the attacker decides to consume the message directly from the client (different from the above three results, where the attacker does not consume the first message from the client to the server). In a scenario where the attacker repeatedly performs such behaviour (i.e. an isolating attacker), we conclude that the protocol for QoS = 1 does not necessarily guarantee a minimum of one delivery of the message to the subscribers, as the protocol specification claims. As the analysis results show, there is a likelihood that the message may never be delivered to the subscribers. Hence, Property 1 cannot be elevated to a theorem as was the case in QoS = 0.

4.3.3 QoS = 2 Protocol

The last protocol represents the highest level of quality of service, indicated by the QoS bit setting of 2. The model of Fig. 4.1 contains again the definitions of the client and the publishing server. Similar to (and for the same reasons as) the case of QoS = 1, we use two channels for the client: c for the first part ending with the sending of *Pubrec* and c' for the second part ending with the sending of *Pubcomp*.

The client process has two parts. The first could be re-iterated, which will result in the *Publish* message being resent with the DUP bit set in case the *Pubrec* message is not received from the server within a time bound of t units. Note here that the standard protocol of [1] is not clear regarding the resent message. There is no explicit mention that the resent publish message is considered different from the original one. The assumption we make is that since DUP is set, then the resent message is a "duplicate" of the original one and therefore it is the same message.

The second part of the client process, *ClientCont*, is instantiated by the first part only if *Pubrec* is received from the server within the time bound t. In this case, it will send a *Pubrel* message to the server parameterised by the same message id as received in the previous message (hence we write $Pubrel_u$). After this, it waits for an amount of time t' for the last message from the server, *Pubcomp*, at which point it terminates once this message is received. If this last message does not arrive within the time bound t', it will re-call itself (i.e. the *ClientCont* part), which will result in the re-commencement of the protocol from the point of the sending of the $Pubrel_u$ message. We believe the above two timed input actions model adequately the requirement *"If a failure is detected, or after a defined time period, the protocol flow is retried from the last unacknowledged protocol message; either the PUBLISH or PUBREL."* [1, pp. 38].

Finally, the last part of the protocol represents the server process. This process after receiving the initial publish message splits into a choice of two processes, *ServerEarly* and *ServerLate*. The main difference between these is whether the publish message is published to the subscribers before or after sending the second message of the protocol $Pubrec_x$, which is parameterised by the message id received in the first message from the client.

The specification provides two alternatives for this case [1, pp. 38]. The first follows the sequence of actions *store message, publish message and delete message*, whereas the second follows the sequence of actions *store message id, publish message and delete message id*. We term the former a *late publish semantics* and the latter an *early publish semantics*. The specification document states that "The choice of semantic is implementation specific and does not affect the guarantees of a QoS level 2 flow" [1, pp. 38], however, we demonstrate next in terms of the output of our static analysis that this is not generally true.

The whole server process is replicated in order to be able to receive a repeat publish message from the client in the event that $Pubrec_x$ is not received at the client within the time limit. The server process, after sending $Pubrec_x$, goes into the second part of the protocol. In this part, it listens on $c'(v)$ or $c'(q)$ for the incoming *Pubrel* message from the client. It then continues depending on the choice made earlier to either publish the message and send $Pubcomp_v$ or just send $Pubcomp_q$. In both cases, the *Pubcomp* message is parameterised by the message id from the received *Pubrel* message.

The final part now commences, which is a replicated process that again listens for the *Pubrel* message from the client, and once this is received, it sends another *Pubcomp* message back to the client. This last part of the server process is similar in both sides of the choice, and it will replicate itself until the client receives successfully the *Pubcomp* message, at which point the client will cease resending *Pubrel* messages.

It is worth noting here that this model assumes that the implementation of the server will cater to a non-deterministic choice of both the early and late publish semantics. However, it is also possible, as we shall see in the next section, to model and analyse the server assuming only one of the two semantics of message publishing is implemented. This would be equivalent to modelling the server process as either $!c(l).ServerLate(l)$ or $!c(l).ServerEarly(l)$.

We now capture the delivery semantics for this protocol in terms of the following property.

Property 4.2 (Delivery Semantics For $QoS = 2$) *The MQTT protocol for the case of QoS = 2 has a delivery semantics of the publish message to the subscribers of "exactly once".*   ☐

In the first analysis we run, the attacker is deactivated. We obtain the following subset value for ϕ for $k = 1$:

$$\phi = \{z_1 \mapsto \{Publish\}, l_1 \mapsto \{Publish\}, x_1 \mapsto \{Publish\}, u_1 \mapsto \{Publish\},$$
$$y_1 \mapsto \{Pubrec_x\}, v_1 \mapsto \{Pubrel_u\}, x_1' \mapsto \{Publish, q_1 \mapsto \{Pubrel_u\}, w_1 \mapsto \{Pubcomp_v,$$
$$Pubcomp_q\}\}$$

The substitutions correspond to normal runs of the protocol for the two choices of the late and early publish semantics, where some values such as for w_1 combine both options. Again, in such normal conditions, increasing k over 1 does not provide any new insight into the protocol delivery semantics.

Next, we examine some of the results of the analysis when the attacker is *activated*. In particular, we consider the case of early publish semantics where we analyse in the context of the server $!c(l).ServerEarly(l)$ (i.e. the server always makes the choice of the $ServerEarly$ process) and the simple attacker model $(c(y') + c'(u'))$. We obtain the following subset of the results, with $k = 2$:

$$\phi_{atk1} = \{x_1 \mapsto \{Publish\}, x_1' \mapsto \{Publish\}, y_1' \mapsto \{Pubrec_x\}, x_2 \mapsto \{Publish_{DUP}\},$$
$$x_2' \mapsto \{Publish_{DUP}\}, y_1 \mapsto \{Pubrec_x\}, q_1 \mapsto \{Pubrel_u\}, w_1 \mapsto \{Pubcomp_q\}, \ldots\}$$

We focus on this interesting subset as it represents a single interference case by the attacker (since $k = 2$). The attacker manages to consume the $Pubrec$ message ($y_1' \mapsto \{Pubrec_x\}$) before the client does so. As a result, the first part of the protocol is repeated and hence, in addition to the initial publish message ($x_1' \mapsto \{Publish\}$), this leads to the second instance of this message to be announced to the subscribers ($x_2' \mapsto \{Publish_{DUP}\}$). The second instance then continues as normal, leading to $y_1 \mapsto \{Pubrec_x\}$.

Next, we re-apply the analysis to the case of the full server model and the simple non-replicated attacker model, where again we set $k = 2$ for simplicity:

$$\phi_{atk2} = \{x_1 \mapsto \{Publish\}, x_1' \mapsto \{Publish\}, y_1' \mapsto \{Pubrec_x\}, x_2 \mapsto \{Publish_{DUP}\},$$
$$y_1 \mapsto \{Pubrec_x\}, v_1 \mapsto \{Pubrel_u\}, x_2' \mapsto \{Publish_{DUP}\}, w_1 \mapsto \{Pubcomp_v\}, \ldots\}$$

This subset of the analysis results represents another case of the attacker interfering with the protocol. Unlike the case of the first attack, a different choice of the publish semantics is made here in terms of the retransmission of the first part of the protocol. Here, we find that the $Pubrec_x$ acknowledgement message sent by the server is captured by the attacker after an early publish semantics choice is taken involving announcing the publish message to the subscribers (by means of $x'_1 \mapsto \{Publish\}$). This failure in delivering $Pubrec_x$ to the client causes a restart of the protocol, however, in this case a different choice is made with the late publish semantics. Continuing with this run, the second part of the protocol causes the duplicated publish message $Publish_{DUP}$ to be announced again to the subscribers. Note that this attack would not be possible if either $ServerLate$ or $ServerEarly$ process only is adopted, but not a choice of both.

These anomalies in the case of QoS $= 2$, particularly those arising from ϕ_{atk2}, were raised as an OASIS issue number MQTT-209 [6] for version. An old paragraph in the specification document [1, pp. 38] stated that "*The choice of semantic is implementation specific and does not affect the guarantees of a QoS level 2 flow.*", and as a result of the above results the latest specification of the protocol has been revised to avoid the ambiguity issue [7, pp. 55]: "*The choice of Method A or Method B is implementation specific. As long as an implementation chooses exactly one of these approaches, this does not affect the guarantees of a QoS 2 flow*".

4.4   Client/Server Timed Input Failures

So far, we have analysed the MQTT protocol under assumptions that both the client and server do not fail, i.e. the loss of messages is entirely a result of communication interferences from the network attacker in its various definitions. Here, we consider the case when both clients and servers fail, and without the presence of the attacker, to analyse the effects of such failures on the protocol.

We use our timed inputs as a mechanism to express failures. More specifically, we define our notion of failure as follows.

Definition 4.1 A failed input is defined as a 0-timed input, $\texttt{timer}^0(x(y).P, Q)$. This failure will cause the input action to never take place leading to P, and instead the alternative process Q will always be chosen.

We limit ourselves in this chapter to such clear failures as in the above definition and avoid *semantic failures*, where for an ideal (correct) time value of t, the input $\texttt{timer}^{t'}(x(y).Q, P)$ is defined such that $t' \neq t$ and $t' > 0$. Such failures require a more in-depth analysis of the requirements and design of the system incorporating the MQTT protocol to determine what the right value for t is.

We consider next the impact of input failures on the QoS cases.

4.4.1 The Case of QoS = 0

For this case, considering the model of Fig. 4.1, the server is the only process that can fail in the above manner since it is the only process that performs an input. We could model this failure in the new server process as follows:

$$Server_{fail} \stackrel{\text{def}}{=} \texttt{timer}^0(c(x).\overline{pub}\langle x \rangle, \mathbf{0})$$

which results in a failed server, which can never accept the initial message from the client containing the data to be published, *Publish*. This failure can be seen as a denial of service (literally) and would be equivalent to the presence of a passive attacker that repeatedly consumes the client's message without relaying it through to the server. Even in the presence of such failure, the semantics of the protocol in the case of QoS = 0 remains intact as specified in [1].

4.4.2 The Case of QoS = 1

In the case of QoS = 1, there are both possibilities that the client and server processes fail. For the case of the client process, t could be set to 0 in the definition of Fig. 4.1 as follows:

$$Client_{fail}(z) \stackrel{\text{def}}{=} \overline{c}\langle z \rangle.\texttt{timer}^0(c'(y), Client_{fail}(Publish_{DUP}))$$

With such failure in the client process, the client ends up iteratively generating the published message without considering the reply from the server. This amounts to a client with "spammy" behaviour, which will have an impact on the server's resources, as the server will repeatedly attempt to reply back to the client's publish messages and possibly keep client connections open, hence draining its resources.

On the other hand, the server itself may incorporate a timed element in its input action that can also fail:

$$Server_{fail}() \stackrel{\text{def}}{=} !\texttt{timer}^0(c(x).\overline{pub}\langle x \rangle.\overline{c'}\langle Puback \rangle, \mathbf{0})$$

This is equivalent to a server that is experiencing downtime (similar to the case of QoS = 0 above), unable to provide any interactions with the clients.

If we have both the client and server experiencing input failures, then the effect will be again a spammy client attempting to publish messages to a failed server. The only impact here would be on the communication infrastructure as the useless messages from the client would consume bandwidth needlessly.

4.4.3 The Case of QoS = 2

This case introduces a complex failure analysis problem into the discussion. This is due to the fact that the number of possible failed subprocesses in the definition of the protocol for the case of QoS = 2 is 4 (i.e. the processes that contain input actions, *Client*, *ClientCont*, *ServerLate* and *ServerEarly*). For the two client processes, *Client* and *ClientCont*, there are four possible definitions:

$$Client_{fail1}(z) \stackrel{\text{def}}{=} \overline{c}\langle z\rangle.\texttt{timer}^0(c(y).ClientCont(y), Client_{fail1}(Publish_{DUP}))$$
$$ClientCont_{fail1}(u) \stackrel{\text{def}}{=} \overline{c'}\langle Pubrel_u\rangle.\texttt{timer}^0(c'(w), ClientCont_{fail1}(u))$$
$$Client_{fail2}(z) \stackrel{\text{def}}{=} \overline{c}\langle z\rangle.\texttt{timer}^0(c(y).ClientCont(y), Client(Publish_{DUP}))$$
$$ClientCont_{fail2}(u) \stackrel{\text{def}}{=} \overline{c'}\langle Pubrel_u\rangle.\texttt{timer}^0(c'(w), ClientCont(u))$$

which are created by setting $t = t' = 0$. The first two processes call internally also failed versions of themselves, whereas the second two definitions call non-failed versions. The call to *ClientCont* from within *Client* has no choices as this call is part of the failed input action meaning it will never take place. On the other hand, the two server subprocesses also have two possible failures for each one of them:

$$ServerLate_{fail1}(x) \stackrel{\text{def}}{=} (\overline{c}\langle Pubrec_x\rangle.\texttt{timer}^0(c'(v).\overline{pub}\langle x\rangle.\overline{c'}\langle Pubcomp_v\rangle.$$
$$!(c'(v').\overline{c'}\langle Pubcomp_{v'}\rangle))))$$
$$ServerLate_{fail2}(x) \stackrel{\text{def}}{=} (\overline{c}\langle Pubrec_x\rangle.c'(v).\overline{pub}\langle x\rangle.\overline{c'}\langle Pubcomp_v\rangle.$$
$$!(\texttt{timer}^0(c'(v').\overline{c'}\langle Pubcomp_{v'}\rangle))))$$
$$ServerEarly_{fail1}(x) \stackrel{\text{def}}{=} (\overline{pub}\langle x\rangle.\overline{c}\langle Pubrec_x\rangle.\texttt{timer}^0(c'(q).\overline{c'}\langle Pubcomp_q\rangle.$$
$$!(c'(q').\overline{c'}\langle Pubcomp_{q'}\rangle))))$$
$$ServerEarly_{fail2}(x) \stackrel{\text{def}}{=} (\overline{pub}\langle x\rangle.\overline{c}\langle Pubrec_x\rangle.c'(q).\overline{c'}\langle Pubcomp_q\rangle.$$
$$!(\texttt{timer}^0(c'(q').\overline{c'}\langle Pubcomp_{q'}\rangle))))$$

These server failure cases represent one of three options for each one, where the last option is redundant and hence not considered:

- *ServerLate* (resp. *ServerEarly*) fails on first input $c'(v)$ (resp. $c'(q)$).
- *ServerLate* (resp. *ServerEarly*) fails on second input $c'(v')$ (resp. $c'(q')$).
- *ServerLate* (resp. *ServerEarly*) fails on both first and second inputs $c'(v)$ and $c'(v')$ (resp. $c'(q)$ and $c'(q')$). This last case is redundant as the failure in the second input is subsumed by the failure in the first input.

Considering all the above failed versions of the client and server processes for the case of QoS = 2, the number of possible combinations of these is $2^8 = 256$, which requires a complex failure analysis.

4.5 Discussion

From the results of the abstract interpretation of all the cases of QoS in the previous sections, we can see that the attacker is always capable of getting hold of the message *Publish* as a result of the attacker being capable of interfering with the communications between the client devices and the servers (brokers). In the case that the published messages are deemed to have some level of secrecy or confidentiality, then the attacker would be compromising this secrecy. In the specification that we considered for the protocol, the channel of communication c is not considered to be private nor encrypted. A private channel would imply some resistance to external interference, e.g. by means of some trusted communication hardware module, and the use of encryption as suggested in the MQTT specification can be handled by means of an SSL-based solution, though this would compromise the lightweight nature of the protocol.

In the case of a more active attacker, such as the following one:

$$Attacker() \stackrel{\text{def}}{=} \ !(c(y') + c'(u')) \ | \ !(\overline{c}\langle msg \rangle) + \overline{c}'\langle msg \rangle)$$

it would be possible to send any locally defined messages, *msg*, in order to disrupt the flow of the protocol and cause incorrect data to be published to the server. More specifically, if $msg \neq Publish$, then the published data will have no integrity. In fact, it is not difficult to see that if $msg = \_\_$, where $\_\_$ is an empty string, then the QoS $= 1$ claim of "at least once delivery" is again undermined from message integrity perspective as the message delivered is always an empty string (despite the fact that there is a message being delivered). This is because running the analysis produces the following result:

$$\phi_{atk} = \{x_1 \mapsto \{\}, u'_1 \mapsto \{\}, x'_1 \mapsto \{\}, y'_1 \mapsto \{Publish\}$$
$$x_2 \mapsto \{\_\_\}, u'_2 \mapsto \{Puback\}, x'_2 \mapsto \{Publish\},$$
$$x_3 \mapsto \{\_\_\}, u'_3 \mapsto \{Puback\}, x'_3 \mapsto \{Publish_{DUP}\},$$
$$x_4 \mapsto \{\_\_\}, u'_4 \mapsto \{Puback\}, x'_4 \mapsto \{Publish_{DUP}\}\}$$

This problem can be solved again by means of concealing the communication channel c or digitally signing messages sent over c to preserve their integrity and source authenticity.

Both of the above two issues, i.e. secrecy and integrity/authenticity of messages, appear in the analysis of Sect. 4.3 at the client–server end of communications. However, this is by no means the only part where this issue may appear. Consider, for example, a further enhancement of the definition of the attacker:

$$Attacker() \stackrel{\text{def}}{=} \ !(c(y') + c'(u') + pub(r)) \ | \ !(\overline{c}\langle msg \rangle) + \overline{c}'\langle msg \rangle + \overline{pub}\langle msg \rangle)$$

In this case, the attacker is also capable of sending and receiving messages on the channel intended for communications between the server and the topic subscribers (i.e. applications) through the use of the *pub* channel. The analysis of the model of MQTT under such an attacker reveals

$$\phi_{atk} = \{x_1 \mapsto \{\}, u'_1 \mapsto \{\}, x'_1 \mapsto \{\}, y'_1 \mapsto \{Publish\}$$
$$x_2 \mapsto \{Publish\}, u'_2 \mapsto \{Puback\}, x'_2 \mapsto \{\_\},$$
$$x_3 \mapsto \{Publish_{DUP}\}, u'_3 \mapsto \{Puback\}, x'_3 \mapsto \{\_\},$$
$$x_4 \mapsto \{Publish_{DUP}\}, u'_4 \mapsto \{Puback\}, x'_4 \mapsto \{\_\}\}$$

where the message sent by the attacker over the *pub* channel is assumed to be the null message $\_$. Again, this channel could be concealed or rendered secure using cryptography if messages exchanged over it are to be kept out of the attacker's sight.

In the case of an attacker that only exercises input capabilities, for example:

$$Attacker() \overset{\text{def}}{=} \ !(c(y') + c'(u') + pub(r))$$

the most pressing issue will be denial of service by means of message losses. It is easy to see that such an attacker is capable of wiping out any communicated messages:

$$\phi_{atk} = \{x_1 \mapsto \{\}, u'_1 \mapsto \{\}, x'_1 \mapsto \{\}, y'_1 \mapsto \{Publish\}$$
$$x_2 \mapsto \{\}, u'_2 \mapsto \{Puback\}, x'_2 \mapsto \{\},$$
$$x_3 \mapsto \{\}, u'_3 \mapsto \{Puback\}, x'_3 \mapsto \{\},$$
$$x_4 \mapsto \{\}, u'_4 \mapsto \{Puback\}, x'_4 \mapsto \{\}\}$$

Such an issue can arise in the specific case of QoS $= 1$, where a passive attacker with replicated input capabilities will undermine the guarantee that the protocol delivers the published messages "at least once" to the relevant topic subscribers. It is worth noting here that this issue can occur again for all QoS levels on the server–subscriber side of communication.

References

1. D. Locke, MQ Telemetry Transport (MQTT) V3.1 Protocol Specification. Technical report, IBM Corporation (2010)
2. L. Zhang, Building facebook messenger (2011). https://www.facebook.com/notes/facebook-engineering/building-facebook-messenger/10150259350998920
3. K. Birman, T. Joseph, SIGOPS Oper. Syst. Rev. **21**(5), 123 (1987)
4. B. Aziz, G. Hamilton, in *Proceedings of the* 2nd *Workshop on the Specification, Analysis and Validation for Emerging Technologies* (Roskilde University, Copenhagen, Denmark, 2002), no. 94 in Datalogiske Skrifter
5. B. Aziz, in *9th International Conference on Availability, Reliability and Security (ARES 2014), Fribourg, Switzerland* (IEEE, 2014)
6. P. Fremantle, QOS 2 delivery options must not be mixed (2014). https://issues.oasis-open.org/browse/MQTT-209
7. A. Banks, R. Gupta, MQTT Version 3.1.1 Plus Errata 01. Technical report, OASIS (2015)

Chapter 5
Second Case Study: The Hermes Protocol

5.1 Introduction

The new wave of digitisation in manufacturing, dubbed Industry 4.0, brings to the manufacturing industry the benefits of many of the recent technological paradigms including Cyber-Physical Systems, Internet of Things, Cloud Computing and Artificial Intelligence. This has led to smarter and more autonomous systems. Of these, the IPC-HERMES-9852 standard [1] has emerged as a leading communication protocol in the electronics manufacturing industry, and as a replacement for the IPC-SMEMA-9851 Mechanical Equipment Interface Standard [2]. The Hermes protocol describes necessary steps in transporting a Printed Circuit Board (PCB) from one machine to another and a description of the scenarios that may occur in the event of errors detected at either the sending or the receiving machine. Hermes is maintained by the Institute for Printed Circuits (IPC) as a non-proprietary open protocol bringing the benefits of Industry 4.0 to the electronics community. The normal operation of the Hermes protocol is illustrated in Fig. 5.1, with alternative sequences of handshakes.

An Upstream machine is a machine that sends a PCB whereas a Downstream machine is a machine that receives the PCB. Note that the protocol provides for two alternatives for two of its message blocks. The first is related to the exchange of the first two messages, MachineReady() and BoardAvailable(), which simply allow each machine to indicate to the other that it is ready to commence the PCB transfer. After that, the Downstream machine starts the transfer by sending the StartTransport() signal. The second alternative sequence then commences, and it is related to the final stage of the transfer, where either machine again can indicate before the other that it has completed the transfer. This is done using the pair of signals TransportFinished(Complete) and StopTransport(Complete) signals. All across the protocol time line, either machine can switch on and switch off its conveyor belt by performing Conveyor on and Conveyor off internal actions, respectively, at the appropriate points in time. Further details of the sequence of messages and their meaning can be found in [1], which also describes a few error-detection scenarios.

© The Author(s), under exclusive license to Springer Nature Switzerland AG 2022   77
B. Aziz, *Formal Analysis by Abstract Interpretation*, SpringerBriefs in Applied
Sciences and Technology, https://doi.org/10.1007/978-3-030-91153-9_5

Fig. 5.1 The normal
operation of Hermes [1] with
alternative sequences

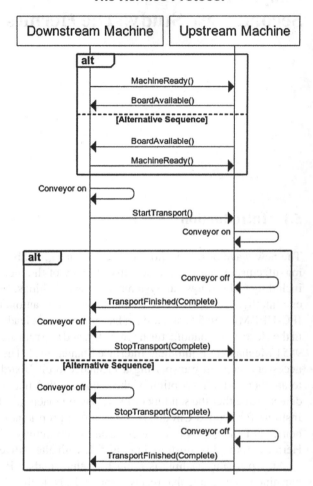

It is worth noting that despite the fact that Hermes is, in its essence, an Industry 4.0 machine-2-machine protocol, it nonetheless provides for connectivity with the Internet, since both the upstream and downstream machines can offer a TCP server that can be contacted on port 1248 for the purpose of receiving configuration messages to configure the machines. This facilitates the deployment of a Hermes-based manufacturing line in an Industrial Internet-of-Things setup.

5.2 A Formal Model of the Hermes Protocol

The formal model for the Hermes protocol reflects the two machines specified in the Hermes protocol [1]: the *upstream* and the *downstream* machines. These are formally defined in Fig. 5.2. This definition reflects the normal operation of these machines without the presence of any errors (i.e. stop signals). The conveyor-on and conveyor-off actions in the Hermes specification document are modelled here for simplicity as silent internal actions, τ. This is due to the fact that these actions cannot be interrupted and therefore do not constitute any externally observable behaviour. Running the full system, *System*, corresponds to running the two machines in parallel:

$$System \stackrel{def}{=} (UP \mid DP)$$

An important property of this protocol is to ensure that both machines agree on whether they have completed the PCB transfer successfully. We express this property as a predicate on the result of analysing the protocol.

Property 5.1 (Agreement on PCB Transfer Completion) *Both machines, upstream and downstream, agree on the PCB transfer completion iff the following predicate, F_1, defined on the result of analysing the protocol, ϕ, is true:*

$$F_1(\phi) = \exists x \in bn(UP), y \in bn(DP):$$
$$\text{"Complete"} \in \phi(x) \Leftrightarrow \text{"Complete"} \in \phi(y) \qquad \square$$

This property captures the requirement that if one machine believes it has completed the transfer of the PCB, the other does as well, in whichever order that completion is signalled. Applying our static analysis, $\mathcal{A}(\!(System)\!)\phi_0 = \phi$, to the protocol specification of Fig. 5.2, we obtain the following two sub-environments:

The Upstream Machine Process:

$$UP \stackrel{def}{=} machineready().\overline{boardavailable}\langle\rangle.UP_{cont} +$$
$$\overline{boardavailable}\langle\rangle.machineready().UP_{cont}$$

where,

$$UP_{cont} \stackrel{def}{=} \overline{starttransport}().\tau.$$
$$(\tau.transportfinished\langle\text{"Complete"}\rangle.\overline{stoptransport}(x).0$$
$$+ stoptransport(x').\tau.\overline{transportfinished}\langle x'\rangle.0)$$

The Downstream Machine Process:

$$DP \stackrel{def}{=} \overline{machineready}\langle\rangle.boardavailable().DP_{cont} +$$
$$boardavailable().\overline{machineready}\langle\rangle.DP_{cont}$$

where,

$$DP_{cont} \stackrel{def}{=} \tau.\overline{starttransport}\langle\rangle.(transportfinished(y').\tau.\overline{stoptransport}\langle y'\rangle.0$$
$$+ \tau.\overline{stoptransport}\langle\text{"Complete"}\rangle.transportfinished(y).0)$$

Fig. 5.2 A π-calculus model of the Hermes v1.2 protocol in normal operation

$$\phi = \phi_1[y' \mapsto \text{``Complete''}, x \mapsto y'] \cup \phi_2[x' \mapsto \text{``Complete''}, y \mapsto x']$$

where both ϕ_1 and ϕ_2 satisfy the above PCB transfer completion predicate F_1. In this case, we did not need to use any approximation function since the set of names used is finite by nature due to the lack of any infinite (replicated) behaviour in the protocol's definition.

5.3   Error Tests

The definition of the Hermes protocol presented in Fig. 5.2 of the previous section assumes normal operation and therefore does not cater to situations when errors are detected by any of the two machines, which could result in stopping their operation. Hence, we define in this section an enhanced formal specification that can handle such transport errors along the lines of the scenarios captured in [1, Sect. 2.3.4].

Before we can do that, we need to define a mechanism in the formal protocol specification to express the ability of the Hermes protocol in testing for the presence of such errors. One such mechanism for defining these *tests* can be specified using the condition evaluation construct in the π-calculus language:

$$state(e_i).P_{e_i}[e_i = \text{``stop''}]Q_{e_i}$$

The communication channel, *state*, is used to receive from the environment of the two machines any "*stop*" signals sent to those machines whenever transport errors occur in them. We use the enumerations, $i \in \mathbb{N}$, to distinguish the different copies of the input parameter, e, that receives such "*stop*" signals from the environment. Process P_{e_i} will run if a "*stop*" signal is received, otherwise, Q_{e_i} runs if the received signal is different from "*stop*". We call P_{e_i} the *error-handling process*.

For brevity, we use the following shorthand representation of the above tests:

$$[e_i](Q_{e_i}) \equiv state(e_i).P_{e_i}[e_i = \text{``stop''}]Q_{e_i}$$

As a matter of fact, $[e_i]$ becomes an *alias* to uniquely reference the communication channel, *state*, input parameter, e_i, and the process, P_{e_i}. We stress here that these tests are part of the operation of the protocol itself, not a method to test the protocol's behaviour.

The next question is *where should such tests be positioned in the protocol's specification?* The Hermes protocol [1, Sect. 2.3.4] provides only a limited number of points at which such tests are carried out (see the seven Scenarios U1a–D3). In our case, we consider that any sequential composition point with an action $a \in \{x(y), \overline{x}\langle y \rangle, \tau\}$ is a potential testing point that the protocol could use to detect errors in machines:

$$\overset{\textit{test point}}{a \overset{\frown}{} P}$$

The Upstream Machine Process:

$UP \overset{\text{def}}{=}$

$machineready().([e_1](\overline{boardavailable}\langle\rangle.([e_2](UP_{cont}) + \overline{revokemachineready}().\mathbf{0})) +$
$\qquad\qquad \overline{revokemachineready}().\mathbf{0}) +$
$\overline{boardavailable}\langle\rangle.[e_3](machineready().([e_4](UP_{cont}) + \overline{revokemachineready}().\mathbf{0}))$

where,

$UP_{cont} \overset{\text{def}}{=} starttransport().[e_5](\tau.[e_6]($
$\qquad\qquad (\tau.[e_7](\overline{transportfinished}\langle"Complete"\rangle.[e_8](stoptransport(x).[e_9](\mathbf{0}))) +$
$\qquad\qquad stoptransport(x').[e_{10}](\tau.[e_{11}](\overline{transportfinished}\langle x'\rangle.[e_{12}](\mathbf{0}))))$

The Downstream Machine Process:

$DP \overset{\text{def}}{=}$

$\overline{machineready}\langle\rangle.[e'_1](boardavailable().([e'_2](DP_{cont}) + revokeboardavailable().\mathbf{0})) +$
$boardavailable().([e'_3](\overline{machineready}\langle\rangle.([e'_4](DP_{cont}) + revokeboardavailable().\mathbf{0}) +$
$\qquad\qquad revokeboardavailable().\mathbf{0}))$

where,

$DP_{cont} \overset{\text{def}}{=} \tau.[e'_5](\overline{starttransport}\langle\rangle.[e'_6]($
$\qquad\qquad (transportfinished(y').[e'_7](\tau.[e'_8](\overline{stoptransport}\langle y'\rangle.[e'_9](\mathbf{0}))) +$
$\qquad\qquad \tau.[e'_{10}](\overline{stoptransport}\langle"Complete"\rangle.[e'_{11}](transportfinished(y).[e'_{12}](\mathbf{0}))))))$

The Test Environment Process:

$Env \overset{\text{def}}{=} (!\,\overline{state}\langle"continue"\rangle.\mathbf{0}) \mid (!\,\overline{state}\langle"stop"\rangle.\mathbf{0})$

Fig. 5.3 A formal model of the Hermes v1.2 protocol with error tests

This implies that each machine should be able to be stopped *after* each action it performs, allowing for a better range of detection of errors at more points in its operation.

Based on this, we redefine the formal specification of the Hermes protocol, as in Fig. 5.3, in order to incorporate the ability to receive stop signals from the environment. The full system, *System*, will now consist of the two machines running in parallel with the environment, *Env*, as follows:

$$System \overset{\text{def}}{=} (UP \mid DP \mid Env)$$

The environment process, *Env*, will always send either a *"continue"* message, when there is no error to be signalled to either machine, or a *"stop"* message, when some transport error is detected. As a result, the definitions of *UP* and *DP* need to be modified, from their normal definitions in Fig. 5.2, to be able to detect *"stop"* messages and to react to any error-handling behaviour such messages may initiate.

The specification of Fig. 5.3 considers all possible points in a process where an error can be tested, once that process has commenced running (i.e. it has fired its first action). We identified 24 such points, compared to the original 7 points that were identified in the scenarios of [1, Sect. 2.3.4]. The error-testing specification also modifies the normal operation specification of Fig. 5.2 by adding extra communication

Process Definition	Description
$P_{e_1} \stackrel{\text{def}}{=} 0$	*No corresponding scenario*
$P_{e_2} \stackrel{\text{def}}{=} \overline{revokeboardavailable}\langle\rangle.0$	*Corresponds to Scenario U1a*
$P_{e_3} \stackrel{\text{def}}{=} P_{e_2}$	*No corresponding scenario*
$P_{e_4} \stackrel{\text{def}}{=} P_{e_2}$	*No corresponding scenario*
$P_{e_5} \stackrel{\text{def}}{=} \overline{transportfinished}\langle\text{``NotStarted''}\rangle.$ $stoptransport(x'').0$	*Corresponds to Scenario U1b*
$P_{e_6} \stackrel{\text{def}}{=} \tau.\overline{transportfinished}\langle\text{``Incomplete''}\rangle.$ $stoptransport(x''').0$	*Corresponds to Scenario U2*
$P_{e_7} \stackrel{\text{def}}{=} \overline{transportfinished}\langle\text{``Incomplete''}\rangle.$ $stoptransport(x'''').0$	*No corresponding scenario*
$P_{e_8} \stackrel{\text{def}}{=} stoptransport(x''''').0$	*Corresponds to Scenario U3*
$P_{e_9} \stackrel{\text{def}}{=} 0$	*No corresponding scenario*
$P_{e_{10}} \stackrel{\text{def}}{=} \tau.\overline{transportfinished}\langle\text{``Complete''}\rangle.0$	*No corresponding scenario*
$P_{e_{11}} \stackrel{\text{def}}{=} \overline{transportfinished}\langle\text{``Complete''}\rangle.0$	*No corresponding scenario*
$P_{e_{12}} \stackrel{\text{def}}{=} 0$	*No corresponding scenario*

Fig. 5.4 Definition of error-handling processes for the upstream machine

channels to cater to the RevokeMachineReady and RevokeBoardAvailable signals resulting from transport errors. Both *UP* and *DP* deviate to their error-handling processes once *"stop"* is received.

The individual error-handling processes, P_{e_i}, are defined in Fig. 5.4 for the upstream machine, and Fig. 5.5 for the downstream machine. We also indicate which of the scenarios in [1] each error-handling process corresponds to, or none if it does not correspond to any. An important observation that one can make here is that while the specification of [1] considers only scenarios where errors occur in either machine, the specification does not consider the possibility that errors may occur simultaneously in both machines. That is, the environment of Fig. 5.3 could *"stop"* both machines at the same time. Due to the execution sequences of *UP* and *DP*, this possibility is limited to the following pairs of tests:

$([e_1], [e_1'])$, $([e_2], [e_2'])$, $([e_3], [e_3'])$, $([e_4], [e_4'])$, $([e_4], [e_5'])$, $([e_2], [e_5'])$, $([e_5], [e_6'])$, $([e_6], [e_6'])$, $([e_7], [e_6'])$, $([e_5], [e_{10}'])$, $([e_6], [e_{10}'])$, $([e_7], [e_{10}'])$, $([e_8], [e_7'])$, $([e_8], [e_8'])$, $([e_9], [e_9'])$, $([e_{10}], [e_{11}'])$, $([e_{11}], [e_{11}'])$, $([e_{12}], [e_{12}'])$

In most of these pairs, nothing will occur as neither point will offer actions that can synchronise with the other point in the pair. However, in a few cases, we find that certain error-handling processes may communicate with each other if both are triggered at the same time, as their actions can synchronise. This is the case for the following four pairs:

$([e_8], [e_7'])$, $([e_8], [e_8'])$, $([e_{10}], [e_{11}'])$, $([e_{11}], [e_{11}'])$

Process Definition	Description
$P_{e_1'} \stackrel{\text{def}}{=} \overline{revokemachineready}\langle\rangle.0$	No corresponding scenario
$P_{e_2'} \stackrel{\text{def}}{=} P_{e_1'}$	Corresponds to Scenario D1
$P_{e_3'} \stackrel{\text{def}}{=} 0$	No corresponding scenario
$P_{e_4'} \stackrel{\text{def}}{=} P_{e_1'}$	No corresponding scenario
$P_{e_5'} \stackrel{\text{def}}{=} \tau.P_{e_1'}$	No corresponding scenario
$P_{e_6'} \stackrel{\text{def}}{=} \tau.\overline{stoptransport}\langle\text{"Incomplete"}\rangle.$ $transportfinished(y'').0$	Corresponds to Scenario D2
$P_{e_7'} \stackrel{\text{def}}{=} \tau.\overline{stoptransport}\langle\text{"Complete"}\rangle.0$	Corresponds to Scenario D3
$P_{e_8'} \stackrel{\text{def}}{=} \overline{stoptransport}\langle\text{"Complete"}\rangle.0$	No corresponding scenario
$P_{e_9'} \stackrel{\text{def}}{=} 0$	No corresponding scenario
$P_{e_{10}'} \stackrel{\text{def}}{=} \overline{stoptransport}\langle\text{"Incomplete"}\rangle.$ $transportfinished(y''').0$	No corresponding scenario
$P_{e_{11}'} \stackrel{\text{def}}{=} transportfinished(y'''').0$	No corresponding scenario
$P_{e_{12}'} \stackrel{\text{def}}{=} 0$	No corresponding scenario

Fig. 5.5 Definition of error-handling processes for the downstream machine

Next, we apply our static analysis, $\mathcal{A}([System])\phi_0 = \phi'$, to the above version of the protocol specification with error tests. As a result, we obtain the following environment:

$\phi' = (\phi_1'[x'' \mapsto y', y' \mapsto \text{"NotStarted"}] \cup \quad \phi_2'[x''' \mapsto y', y' \mapsto \text{"Incomplete"}] \cup$
$\phi_3'[x'''' \mapsto y', y' \mapsto \text{"Incomplete"}] \cup \quad \phi_4'[x' \mapsto \text{"Incomplete"}, y'' \mapsto x'] \cup$
$\phi_5'[x' \mapsto \text{"Incomplete"}, y''' \mapsto x'] \cup \quad \phi_6'[x \mapsto y', y' \mapsto \text{"Complete"}] \cup$
$\phi_7'[x''''' \mapsto y', y' \mapsto \text{"Complete"}] \cup \quad \phi_8'[y \mapsto x', x' \mapsto \text{"Complete"}] \cup$
$\phi_9'[y'''' \mapsto x', x' \mapsto \text{"Complete"}] \cup \quad \phi_{10}'[x''''' \mapsto \text{"Complete"}, y' \mapsto$
$\text{"Complete"}] \cup \quad \phi_{11}'[y'''' \mapsto \text{"Complete"}, x' \mapsto \text{"Complete"}] \cup \quad \phi_{12}'[x \mapsto$
$\text{"Complete"}, y' \mapsto \text{"Complete"}])$

Each sub-environment part of ϕ' represents a different path of execution, depending sometimes on which error points are triggered. This brings us to the following two properties.

Property 5.2 (Agreement on PCB Transfer Not Started) *Both machines, upstream and downstream, agree that the PCB transfer has not started iff F_2 is true on the result of the analysis of the protocol, ϕ', where*

$F_2(\phi') = \exists x \in bn(UP), y \in bn(DP): \quad \text{"Not Started"} \in \phi'(x) \Leftrightarrow \text{"Not Started"}$
$\in \phi'(y)$ □

F_2 is satisfied by the ϕ_1' sub-environment. The F_2 predicate corresponds to error-handling in Scenario U1b in [1], where the error occurs in the upstream machine before it starts transporting the PCB. On the other hand, the next property captures all errors that occur in both machines while the PCB is in the middle of being transported.

Property 5.3 (Agreement on PCB Transfer Not Completed) *Both machines, upstream and downstream, agree that the PCB transfer has not been completed iff F_3 is true, where*
$$F_3(\phi') = \exists x \in bn(UP), y \in bn(DP):$$
$$\text{``Incomplete''} \in \phi'(x) \Leftrightarrow \text{``Incomplete''} \in \phi'(y) \qquad \square$$

F_3 is satisfied by $\phi'_2 - \phi'_5$ sub-environments. Of these, ϕ'_2 corresponds to Scenario U2 and ϕ'_4 corresponds to Scenario D2 in [1]. Both ϕ'_3 and ϕ'_5 have no corresponding scenarios and are, therefore, a new result of this analysis.

It is also worth noting now that the PCB transfer completion is satisfied not only due to the F_1 predicate, which was satisfied by the $\phi_1 = \phi'_6$ and $\phi_2 = \phi'_8$ environments, but also due to the additional cases of ϕ'_7, ϕ'_9, ϕ'_{10}, ϕ'_{11} and ϕ'_{12}. In particular, the cases of ϕ'_{10} and ϕ'_{11} are a result of the communications between the pairs $([e_8], [e'_7])$, $([e_8], [e'_8])$, $([e_{10}], [e'_{11}])$ and $([e_{11}][e'_{11}])$ as indicated earlier. The case of ϕ'_7 corresponds to Scenario U3 and the case of ϕ'_{12} to Scenario D3 [1]. However, ϕ'_9 is a new case that does not correspond to any of the standard scenarios. We now define *robustness* as the result that the two machines always agree on the same status in any environment.

Theorem 5.1 (Robustness of Hermes) *The Hermes protocol is robust since $F_1 \wedge F_2 \wedge F_3$.*

Proof sketch. $F_1 \wedge F_2 \wedge F_3$ can be shown to be true from Properties 1–3. $\square$

Our second case study analysis demonstrates two important aspects: We summarise our two major findings as follows: First, specifying the protocol formally has shown that a number of error-detection scenarios were missed by the standard's specification document (in this case, Hermes [1]). The Hermes document identified only 7 such scenarios, whereas we found that there were 24 scenarios that needed consideration. The significance of this finding is that it increases, for any future implementations of the standard, the robustness of testing suites applied to those implementations by increasing the coverage of these tests from 7 cases to 24. Second, performing the formal modelling and analysis exercise of the standard revealed that its informal specification did not consider simultaneous occurrences of errors—only that a single machine detects an error at any single point in time. This is indicated for each and every one of the 7 scenarios highlighted in the document. However, in reality, such errors may occur simultaneously, and therefore it is necessary also to consider the simultaneous triggering of the error-handling processes at both the sending and receiving machines. This could lead to these processes interacting with one another.

References

1. T.H.S. Initiative, IPC-HERMES-9852: The global standard for machine-to-machine communication in SMT assembly (version 1.2). Technical report, IPC (2019)
2. IPC, IPC-SMEMA-9851: Mechanical Equipment Interface Standard. Technical report, IPC - Association Connecting Electronics Industries (2007)

Ref erences

Chapter 6
Third Case Study: An Electric Vehicle Charging Protocol

6.1 Introduction

At the core of most global energy strategies nowadays is the goal of developing smart infrastructures that can harness clean energy ecosystems in order to lower or eliminate the impact of fossil fuels on the environment. To achieve this goal, standardisation efforts have been on the upbeat in recent years aiming at regulating Electric Vehicle (EV) systems communications and interaction with their infrastructures. More specifically, many standards related to the charging of EVs have emerged, for example, [1–3]. However, like any other critical infrastructure, the resilience of EV infrastructures remains at the forefront of the challenges facing the adoption of the new EV technology [4]. In particular, the reliability of the communication protocols used in these infrastructures in the face of failures resulting from bad or ambiguous specifications is an important aspect that contributes to their resilience, and mutation testing and analysis [5] is one technique widely adopted by industry through which this reliability can be understood.

With the rising popularity of EVs, recently, industry moved on to define the Open Charge Point Protocol (OCPP) v.2.0.1. [1] standard for establishing a charging infrastructure for EVs. OCPP is maintained by the Open Charge Alliance (OCA) [6], a global consortium of public and private electric vehicle infrastructure leaders promoting open standards for electric vehicles and their infrastructure.

The architecture of an OCPP infrastructure (see [7, p. 4]) consists of any number of Charging Stations, each referring to any physical system where an electric vehicle can charge itself. These stations are managed by a single Charging Station Management System (CSMS), which also has the information for authorising users for using the Charging Stations under its management. The Charging Station itself may have multiple units of Electric Vehicle Supply Equipment (EVSE), where each EVSE can have multiple Connectors, though only one Connector can be active at any one point in time.

OCPP is not a single protocol, but itself consists of a number of sub-protocols, each implementing one of its use cases. In total, there are 119 such use cases. The focus of this chapter is on one specific critical sub-protocol, namely the Heartbeat Protocol, defined in Use Case G02 [1, p. 185]. This sub-protocol will form the basis

B. Aziz, *Formal Analysis by Abstract Interpretation*, SpringerBriefs in Applied Sciences and Technology, https://doi.org/10.1007/978-3-030-91153-9_6

of our case study presented later in Sect. 6.2. Nonetheless, any other protocol that
can be specified in our formal language (including other OCPP sub-protocols) can
also be subject to the same mutation function defined in the next section.

6.2 The OCPP Heartbeat Protocol

Use Case G02 [1, p. 185] defines a heartbeat protocol between the CSMS and a
Charging Station in the OCPP standard. Like any other heartbeat protocol, the OCPP
heartbeat protocol has the purpose of ensuring that the Charging Station can inform
the CSMS, every now and then, of the fact that it is still alive. The Charging Station
does that by sending regular heartbeat signals to the CSMS. The sequence diagram
for the use case is defined by the following two-way handshake:

$$Charging\ Station \longrightarrow CSMS : HeartbeatRequest()$$
$$CSMS \longrightarrow Charging\ Station : HeartbeatResponse(currentTime)$$

The informal description of the scenario, quoted from [1, p. 185], is as follows:

1. *"If there is no activity for a certain time, the Charging Station sends HeartbeatRequest for ensuring that the CSMS knows that a Charging Station is still
   alive."*
2. *"Upon receipt of HeartbeatRequest, the CSMS responds with HeartbeatResponse.
   The response message contains the current time of the CSMS, which the Charging
   Station MAY use to synchronize its internal clock."*

The use case also states 7 requirements on the protocol implementation, which can
be referred to in [1, p. 185–186]. In order for us to study the mutation effects on the
OCPP Heartbeat protocol, the first formal specifications of the behaviour of a single
Charging Station and its CSMS in our formal language are given. This specification
is shown in Fig. 6.1.

At the top level, the protocol process, *Heartbeat*, is defined in terms of two sub-
protocol processes, *Charging Station* and *CSMS*, composed in parallel with each
other. In addition to these, there are two signals (in the form of output actions), also
composed in parallel with the *Charging Station* and the *CSMS* processes. The first of
these signals, *ChargingStation*⟨*zerotime*⟩, bootstraps the *Charging Station* process,
while at the same time passing it the value of the initial time, referred to as *zerotime*.
It is worth pointing out here that these *time* messages utilised in the protocol are a
different concept from our *time values*, used to define the behaviour of a timed input
action. The second signal, $\overline{CSMSc}$⟨⟩, bootstraps the *CSMS* process. For clarity, these
two signals from now onward are referred to as the *top-level signals*. Note that both
the *ChargingStation* and *CSMSc* channel names are restricted at this level, in order
to ensure that no interference will occur across multiple heartbeat protocol instances
running among multiple Charging Stations and CSMS instances.

$Heartbeat \stackrel{\text{def}}{=}$　　　$(\nu ChargingStation)(\nu CSMSc)($

$\qquad\qquad Charging\ Station\ \mid\ \overline{ChargingStation\langle zeroTime\rangle}.\mathbf{0}\ \mid\ CSMS\ \mid$

$\qquad\qquad \overline{CSMSc\langle\rangle}.\mathbf{0})$

$Charging\ Station \stackrel{\text{def}}{=}$　$!\ \mathtt{timer}^{\infty}(ChargingStation(time).$

$\qquad\qquad \mathtt{timer}^{HeartbeatInterval}(\overline{event()}.ChargingStation\langle time\rangle.\mathbf{0},$

$\qquad\qquad \overline{HeartbeatRequest\langle\rangle}.\mathtt{timer}^{t}(HeartbeatResponse(xt).$

$\qquad\qquad \overline{ChargingStation\langle xt\rangle}.\mathbf{0},\ \overline{ChargingStation\langle time\rangle}.\mathbf{0})),\mathbf{0})$

$CSMS \stackrel{\text{def}}{=}$　　　$!\ \mathtt{timer}^{\infty}(CSMSc().\mathtt{timer}^{\infty}(HeartbeatRequest().$

$\qquad\qquad \overline{HeartbeatResponse\langle currentTime\rangle}.\overline{CSMSc\langle\rangle}.\mathbf{0},\mathbf{0}),\mathbf{0})$

Fig. 6.1 Formal specification of the OCPP heartbeat protocol

The *Charging Station* process itself is a replicated process, which allows for any number of times for its behaviour to be run. This behaviour is a timed input action that waits forever to receive a time value. After receiving its initial time value from the top-level signal, it waits to receive an external event from the *CSMS* process, not related to the Heartbeat protocol. This second input is timed with the value of the *HeartbeatInterval* variable, stated in the use case, and it is performed over a hypothetical channel called *event*. If such activity event is received, the *Charging Station* will then start a new copy of itself, passing itself the current time value message it had initially received. If, however, no such event is received, the actual heartbeat sequence of messages kicks in. This sequence of messages starts by sending a heartbeat request, and then waiting for t amount of time to receive the heartbeat response from the *CSMS* process. If such a response is received, it will carry with it the current time message, which is then used to restart a new copy of the *Charging Station* process. If, however, it is not received, the existing time message received from the above top-level signal is used to start the new copy of *Charging Station*. The specification of the Heartbeat protocol in [1, p. 185] does not specify what the value of t should be. In other words, the document does not define the time limit that the Charging Station should wait for, between sending a heartbeat request and receiving the corresponding heartbeat response from CSMS.

The *CSMS* replicated process, on the other hand, has simpler behaviour. It will wait infinitely to be initiated by the top-level signal, then it will wait again infinitely for a heartbeat request signal to arrive from *Charging Station*. When such a signal is received, it will send the *currentTime* message back to the *Charging Station*, and restart a new copy of itself.

In all of the above definitions, whenever an alternative to an expired timed input is not specified, it is assumed to be the inactive process, $\mathbf{0}$. This is because the use case itself does not state such alternative behaviour, and therefore, it will not form a part of the analysis presented here.

6.3 Analysis of the Heartbeat Protocol

The starting point here is analysing the Heartbeat protocol itself without any mutations applied. Therefore, the analysis, $\mathcal{A}(\![Heartbeat]\!)\phi_0 = \phi_{normal}$, is performed, where $\phi_{normal} = \phi' \cup_\phi \phi''$, and where

$\phi'[time_1 \mapsto zeroTime, time_2 \mapsto time_1, time_2 \mapsto time_2]$
$\phi''[time_1 \mapsto zeroTime, xt_1 \mapsto currentTime, time_2 \mapsto xt_1, xt_2 \mapsto currentTime,$
    $time_2 \mapsto xt_2]$

6.4 A Mutation Framework

In general terms, mutation testing and analysis is concerned with the concept of introducing small single faults or errors into a system's design, specification, implementation or interface and then testing or analysing the effects caused by those faults or errors. It has also been used as a technique for evaluating the quality of test suites, as the more faults the tests can spot, the better the quality of those tests is. In some sense, mutating systems simulate real-world scenarios, where the behaviour of the system is not well-defined or well-implemented, and its communicated messages can be subject to unexpected alterations either by the physical nature of the communication medium, such as any lossy or harsh network, or by some intended or unintended interference from users or other systems, for example, due to actions carried out by malicious external intruders.

Yet, another source of such faults, or mutations, is the decisions made at the specification level, or implementation decisions that we can model back to the initial specification level. Such "high-level" mutations require studying and inspection at the level of a system's specification, since their effect could be more fundamental than mutations at the more technical levels of the software development life cycle. The problem becomes even more urgent when dealing with specifications of standards, since standards are implemented in many ways and could have a far more reaching impact on technologies than any one specific system.

We start the definition of our mutation framework by first defining *tags* that can be used to tag channel and message names as well as time values, i.e. the elements that we will mutate later on. More formally, such tags are defined as the set $\mathcal{T}$ ranged over by elements ℓ, ℓ', etc. We refer to the set of all *tagged processes* as $\mathcal{P}^\mathcal{T}$. A tagged process can be obtained from a normal process through the application of the tagging functions defined in Sect. 6.4.1, which will later be used when generating our mutants.

Formally, the syntax of a tagged process is defined as follows:

$$P, Q ::= \overline{\hat{x}}\langle\hat{y}\rangle.P \mid \texttt{timer}^{\hat{t}}(\hat{x}(y).P, Q) \mid {!P} \mid (\nu x)P \mid (P \mid Q) \mid (P+Q) \mid \mathbf{0}$$

where names, $\hat{x}$, $\hat{y}$, can either be non-tagged or tagged (again allowing for the case when no names are passed as messages). We call the set of all tagged names, $\mathcal{N}^\mathcal{T}$. Similarly, time values, $\hat{t}$, can be either natural numbers and therefore elements of $\mathbb{N}$, or tagged natural numbers, therefore elements of the set $\mathbb{N}^\mathcal{T}$. In all that follows, we assume that every tag is unique. This can always be achieved through α-renaming of tags. Tagging has no effect on the structural operational semantics of a process.

6.4.1 Three Tagging Functions

As we mentioned earlier, our focus in this chapter is to study the effects of three kinds of mutations: mutations of messages, communication channels and of the duration of input actions. We define three specific tagging functions to assist in generating mutants for each of these three cases:

$$\gamma_{chn} : \mathcal{P}^\mathcal{T} \to \mathcal{P}^\mathcal{T} \tag{6.1}$$

$$\gamma_{msg} : \mathcal{P}^\mathcal{T} \to \mathcal{P}^\mathcal{T} \tag{6.2}$$

$$\gamma_{time} : \mathcal{P}^\mathcal{T} \to \mathcal{P}^\mathcal{T} \tag{6.3}$$

As their names imply, γ_{chn} tags channel names, γ_{msg} tags message names and γ_{time} tags the time duration of an input action. These will simplify and generalise the definition of the mutation function, introduced later in Sect. 6.4.3. The syntax-directed rules defining these functions are shown in Figs. 6.2, 6.3 and 6.4, respectively. The operator

$$tags : \mathcal{P}^\mathcal{T} \to \wp(\mathcal{T}) \tag{6.4}$$

collects the set of tags of a process, as follows:

$(\gamma_{chn}1)$ $\gamma_{chn}([\overline{x}\langle \hat{y}\rangle.P])\rho$ $\qquad = \overline{x}^{\ell}\langle \hat{y}\rangle.P \quad where\ \ell \notin (\rho \cup tags(\overline{x}\langle \hat{y}\rangle.P))$

$(\gamma_{chn}2)$ $\gamma_{chn}([\overline{x}\langle \rangle.P])\rho$ $\qquad = \overline{x}^{\ell}\langle \rangle.P \quad where\ \ell \notin (\rho \cup tags(\overline{x}\langle \rangle.P))$

$(\gamma_{chn}3)$ $\gamma_{chn}([\overline{x}^{\ell}\langle \hat{y}\rangle.P])\rho$ $\qquad = \overline{x}^{\ell}\langle \hat{y}\rangle.\gamma_{chn}([P])(\rho \cup tags(\overline{x}^{\ell}\langle \hat{y}\rangle.0))$

$(\gamma_{chn}4)$ $\gamma_{chn}([\overline{x}^{\ell}\langle \rangle.P])\rho$ $\qquad = \overline{x}^{\ell}\langle \rangle.\gamma_{chn}([P])(\rho \cup tags(\overline{x}^{\ell}\langle \rangle.0))$

$(\gamma_{chn}5)$ $\gamma_{chn}([\mathtt{timer}^{\hat{t}}(x(y).P,Q)])\rho = \mathtt{timer}^{\hat{t}}(x^{\ell}(y).P,Q)$
$\qquad\qquad\qquad\qquad\qquad\qquad\qquad where\ \ell \notin (\rho \cup tags(\mathtt{timer}^{\hat{t}}(x(y).P,Q)))$

$(\gamma_{chn}6)$ $\gamma_{chn}([\mathtt{timer}^{\hat{t}}(x().P,Q)])\rho = \mathtt{timer}^{\hat{t}}(x^{\ell}().P,Q)$
$\qquad\qquad\qquad\qquad\qquad\qquad\qquad where\ \ell \notin (\rho \cup tags(\mathtt{timer}^{\hat{t}}(x().P,Q)))$

$(\gamma_{chn}7)$ $\gamma_{chn}([\mathtt{timer}^{\hat{t}}(x^{\ell}(y).P,Q)])\rho =$

$$
\begin{cases}
\mathtt{timer}^{\hat{t}}(x^{\ell}(y).\gamma_{chn}([P])\rho',Q) & if\ tags(\gamma_{chn}([P])\rho')\backslash tags(P) \neq \{\} \\
\mathtt{timer}^{\hat{t}}(x^{\ell}(y).P,\gamma_{chn}([Q])\rho'') & if\ (tags(\gamma_{chn}([P])\rho')\backslash tags(P) = \{\}) \wedge \\
& \quad (tags(\gamma_{chn}([Q])\rho'')\backslash tags(Q) \neq \{\}) \\
\mathtt{timer}^{\hat{t}}(x^{\ell}(y).P,Q) & if\ (tags(\gamma_{chn}([P])\rho')\backslash tags(P) = \{\}) \wedge \\
& \quad (tags(\gamma_{chn}([Q])\rho'')\backslash tags(Q) = \{\})
\end{cases}
$$
$$\qquad\qquad where\ \rho' = \rho \cup tags(\mathtt{timer}^{\hat{t}}(x^{\ell}(y).0,Q))$$
$$\qquad\qquad and\ \rho'' = \rho \cup tags(\mathtt{timer}^{\hat{t}}(x^{\ell}(y).P,0))$$

$(\gamma_{chn}8)$ $\gamma_{chn}([\mathtt{timer}^{\hat{t}}(x^{\ell}().P,Q)])\rho =$

$$
\begin{cases}
\mathtt{timer}^{\hat{t}}(x^{\ell}().\gamma_{chn}([P])\rho',Q) & if\ tags(\gamma_{chn}([P])\rho')\backslash tags(P) \neq \{\} \\
\mathtt{timer}^{\hat{t}}(x^{\ell}().P,\gamma_{chn}([Q])\rho'') & if\ (tags(\gamma_{chn}([P])\rho')\backslash tags(P) = \{\}) \wedge \\
& \quad (tags(\gamma_{chn}([Q])\rho'')\backslash tags(Q) \neq \{\}) \\
\mathtt{timer}^{\hat{t}}(x^{\ell}().P,Q) & if\ (tags(\gamma_{chn}([P])\rho')\backslash tags(P) = \{\}) \wedge \\
& \quad (tags(\gamma_{chn}([Q])\rho'')\backslash tags(Q) = \{\})
\end{cases}
$$
$$\qquad\qquad where\ \rho' = \rho \cup tags(\mathtt{timer}^{\hat{t}}(x^{\ell}().0,Q))$$
$$\qquad\qquad and\ \rho'' = \rho \cup tags(\mathtt{timer}^{\hat{t}}(x^{\ell}().P,0))$$

$(\gamma_{chn}9)$ $\gamma_{chn}([P \mid Q])\rho =$

$$
\begin{cases}
\gamma_{chn}([P])\rho' \mid Q & if\ tags(\gamma_{chn}([P])\rho')\backslash tags(P) \neq \{\} \\
P \mid \gamma_{chn}([Q])\rho'' & if\ (tags(\gamma_{chn}([P])\rho')\backslash tags(P) = \{\}) \wedge \\
& \quad (tags(\gamma_{chn}([Q])\rho'')\backslash tags(Q) \neq \{\}) \\
P \mid Q & if\ (tags(\gamma_{chn}([P])\rho')\backslash tags(P) = \{\}) \wedge \\
& \quad (tags(\gamma_{chn}([Q])\rho'')\backslash tags(Q) = \{\})
\end{cases}
$$
$$\qquad\qquad where\ \rho' = \rho \cup tags(Q)$$
$$\qquad\qquad and\ \rho'' = \rho \cup tags(P)$$

$(\gamma_{chn}10)$ $\gamma_{chn}([!P])\rho$ $\qquad\qquad = !\gamma_{chn}([P])\rho$

$(\gamma_{chn}11)$ $\gamma_{chn}([(\nu x)P])\rho$ $\qquad\quad = (\nu x)\gamma_{chn}([P])\rho$

$(\gamma_{chn}12)$ $\gamma_{chn}([0])\rho$ $\qquad\qquad\quad = 0$

$(\gamma_{chn}13)$ $\gamma_{chn}([P+Q])\rho =$

$$
\begin{cases}
\gamma_{chn}([P])\rho' + Q & if\ tags(\gamma_{chn}([P])\rho')\backslash tags(P) \neq \{\} \\
P + \gamma_{chn}([Q])\rho'' & if\ (tags(\gamma_{chn}([P])\rho')\backslash tags(P) = \{\}) \wedge \\
& \quad (tags(\gamma_{chn}([Q])\rho'')\backslash tags(Q) \neq \{\}) \\
P + Q & if\ (tags(\gamma_{chn}([P])\rho')\backslash tags(P) = \{\}) \wedge \\
& \quad (tags(\gamma_{chn}([Q])\rho'')\backslash tags(Q) = \{\})
\end{cases}
$$
$$\qquad\qquad where\ \rho' = \rho \cup tags(Q)$$
$$\qquad\qquad and\ \rho'' = \rho \cup tags(P)$$

Fig. 6.2 Definition of the channel tagging function

$(\gamma_{msg}1)$ $\gamma_{msg}(\llbracket\bar{\hat{x}}\langle y\rangle.P\rrbracket)\rho$ $=\bar{\hat{x}}\langle y^\ell\rangle.P$ where $\ell\notin(\rho\cup tags(\bar{\hat{x}}\langle y\rangle.P))$

$(\gamma_{msg}2)$ $\gamma_{msg}(\llbracket\bar{\hat{x}}\langle y^\ell\rangle.P\rrbracket)\rho$ $=\bar{\hat{x}}\langle y^\ell\rangle.\gamma_{msg}(\llbracket P\rrbracket)(\rho\cup tags(\bar{\hat{x}}\langle y^\ell\rangle.0))$

$(\gamma_{msg}3)$ $\gamma_{msg}(\llbracket\bar{\hat{x}}\langle\rangle.P\rrbracket)\rho$ $=\bar{\hat{x}}\langle\rangle.\gamma_{msg}(\llbracket P\rrbracket)(\rho\cup tags(\hat{x}\langle\rangle.0))$

$(\gamma_{msg}4)$ $\gamma_{msg}(\llbracket\texttt{timer}^f(\hat{x}(y).P,Q)\rrbracket)\rho=$

$\begin{cases}\texttt{timer}^f(\hat{x}(y).\gamma_{msg}(\llbracket P\rrbracket)\rho',Q) & \text{if } tags(\gamma_{msg}(\llbracket P\rrbracket)\rho')\backslash tags(P)\neq\{\}\\[4pt]\texttt{timer}^f(\hat{x}(y).P,\gamma_{msg}(\llbracket Q\rrbracket)\rho'') & \text{if }(tags(\gamma_{msg}(\llbracket P\rrbracket)\rho')\backslash tags(P)=\{\})\wedge\\ & \qquad(tags(\gamma_{msg}(\llbracket Q\rrbracket)\rho'')\backslash tags(Q)\neq\{\})\\[4pt]\texttt{timer}^f(\hat{x}(y).P,Q) & \text{if }(tags(\gamma_{msg}(\llbracket P\rrbracket)\rho')\backslash tags(P)=\{\})\wedge\\ & \qquad(tags(\gamma_{msg}(\llbracket Q\rrbracket)\rho'')\backslash tags(Q)=\{\})\\ & \text{where }\rho'=\rho\cup tags(\texttt{timer}^f(\hat{x}(y).0,Q))\\ & \text{and }\rho''=\rho\cup tags(\texttt{timer}^f(\hat{x}(y).P,0))\end{cases}$

$(\gamma_{msg}5)$ $\gamma_{msg}(\llbracket\texttt{timer}^f(\hat{x}().P,Q)\rrbracket)\rho=$

$\begin{cases}\texttt{timer}^f(\hat{x}().\gamma_{msg}(\llbracket P\rrbracket)\rho',Q) & \text{if } tags(\gamma_{msg}(\llbracket P\rrbracket)\rho')\backslash tags(P)\neq\{\}\\[4pt]\texttt{timer}^f(\hat{x}().P,\gamma_{msg}(\llbracket Q\rrbracket)\rho'') & \text{if }(tags(\gamma_{msg}(\llbracket P\rrbracket)\rho')\backslash tags(P)=\{\})\wedge\\ & \qquad(tags(\gamma_{msg}(\llbracket Q\rrbracket)\rho'')\backslash tags(Q)\neq\{\})\\[4pt]\texttt{timer}^f(\hat{x}().P,Q) & \text{if }(tags(\gamma_{msg}(\llbracket P\rrbracket)\rho')\backslash tags(P)=\{\})\wedge\\ & \qquad(tags(\gamma_{msg}(\llbracket Q\rrbracket)\rho'')\backslash tags(Q)=\{\})\\ & \text{where }\rho'=\rho\cup tags(\texttt{timer}^f(\hat{x}().0,Q))\\ & \text{and }\rho''=\rho\cup tags(\texttt{timer}^f(\hat{x}().P,0))\end{cases}$

$(\gamma_{msg}6)$ $\gamma_{msg}(\llbracket P\mid Q\rrbracket)\rho=$

$\begin{cases}\gamma_{msg}(\llbracket P\rrbracket)\rho'\mid Q & \text{if } tags(\gamma_{msg}(\llbracket P\rrbracket)\rho')\backslash tags(P)\neq\{\}\\[4pt]P\mid\gamma_{msg}(\llbracket Q\rrbracket)\rho'' & \text{if }(tags(\gamma_{msg}(\llbracket P\rrbracket)\rho')\backslash tags(P)=\{\})\wedge\\ & \qquad(tags(\gamma_{msg}(\llbracket Q\rrbracket)\rho'')\backslash tags(Q)\neq\{\})\\[4pt]P\mid Q & \text{if }(tags(\gamma_{msg}(\llbracket P\rrbracket)\rho')\backslash tags(P)=\{\})\wedge\\ & \qquad(tags(\gamma_{msg}(\llbracket Q\rrbracket)\rho'')\backslash tags(Q)=\{\})\\ & \text{where }\rho'=\rho\cup tags(Q)\\ & \text{and }\rho''=\rho\cup tags(P)\end{cases}$

$(\gamma_{msg}7)$ $\gamma_{msg}(\llbracket!P\rrbracket)\rho$ $=!\gamma_{msg}(\llbracket P\rrbracket)\rho$

$(\gamma_{msg}8)$ $\gamma_{msg}(\llbracket(\nu x)P\rrbracket)\rho$ $=(\nu x)\gamma_{msg}(\llbracket P\rrbracket)\rho$

$(\gamma_{msg}9)$ $\gamma_{msg}(\llbracket 0\rrbracket)\rho$ $=0$

$(\gamma_{msg}10)$ $\gamma_{msg}(\llbracket P+Q\rrbracket)\rho=$

$\begin{cases}\gamma_{msg}(\llbracket P\rrbracket)\rho'+Q & \text{if } tags(\gamma_{msg}(\llbracket P\rrbracket)\rho')\backslash tags(P)\neq\{\}\\[4pt]P+\gamma_{msg}(\llbracket Q\rrbracket)\rho'' & \text{if }(tags(\gamma_{msg}(\llbracket P\rrbracket)\rho')\backslash tags(P)=\{\})\wedge\\ & \qquad(tags(\gamma_{msg}(\llbracket Q\rrbracket)\rho'')\backslash tags(Q)\neq\{\})\\[4pt]P+Q & \text{if }(tags(\gamma_{msg}(\llbracket P\rrbracket)\rho')\backslash tags(P)=\{\})\wedge\\ & \qquad(tags(\gamma_{msg}(\llbracket Q\rrbracket)\rho'')\backslash tags(Q)=\{\})\\ & \text{where }\rho'=\rho\cup tags(Q)\\ & \text{and }\rho''=\rho\cup tags(P)\end{cases}$

Fig. 6.3 Definition of the message-tagging function

$(\gamma_{time}1) \quad \gamma_{time}([\![\overline{\hat{x}}\langle\hat{y}\rangle.P]\!])\rho \qquad\qquad = \overline{\hat{x}}\langle\hat{y}\rangle.\gamma_{time}([\![P]\!])(\rho \cup tags(\overline{\hat{x}}\langle\hat{y}\rangle.\mathbf{0}))$

$(\gamma_{time}2) \quad \gamma_{time}([\![\overline{\hat{x}}\langle\rangle.P]\!])\rho \qquad\qquad = \overline{\hat{x}}\langle\rangle.\gamma_{time}([\![P]\!])(\rho \cup tags(\overline{\hat{x}}\langle\rangle.\mathbf{0}))$

$(\gamma_{time}3) \quad \gamma_{time}([\![\texttt{timer}^{t}(\hat{x}(y).P,Q)]\!])\rho = \texttt{timer}^{t^{\ell}}(\hat{x}(y).P,Q)$
$\qquad\qquad\qquad\qquad\qquad\qquad\qquad where\ \ell \notin (\rho \cup tags(\texttt{timer}^{t}(\hat{x}(y).P,Q)))$

$(\gamma_{time}4) \quad \gamma_{time}([\![\texttt{timer}^{t}(\hat{x}().P,Q)]\!])\rho = \texttt{timer}^{t^{\ell}}(\hat{x}().P,Q)$
$\qquad\qquad\qquad\qquad\qquad\qquad\qquad where\ \ell \notin (\rho \cup tags(\texttt{timer}^{t}(\hat{x}().P,Q)))$

$(\gamma_{time}5) \quad \gamma_{time}([\![\texttt{timer}^{t^{\ell}}(\hat{x}(y).P,Q)]\!])\rho =$
$\left\{\begin{array}{ll} \texttt{timer}^{\ell}(\hat{x}(y).\gamma_{time}([\![P]\!])\rho',Q) & if\ tags(\gamma_{time}([\![P]\!])\rho')\backslash tags(P) \neq \{\} \\ \texttt{timer}^{\ell}(\hat{x}(y).P,\gamma_{time}([\![Q]\!])\rho'') & if\ (tags(\gamma_{time}([\![P]\!])\rho')\backslash tags(P) = \{\}) \wedge \\ & \quad (tags(\gamma_{time}([\![Q]\!])\rho'')\backslash tags(Q) \neq \{\}) \\ \texttt{timer}^{\ell}(\hat{x}(y).P,Q) & if\ (tags(\gamma_{time}([\![P]\!])\rho')\backslash tags(P) = \{\}) \wedge \\ & \quad (tags(\gamma_{time}([\![Q]\!])\rho'')\backslash tags(Q) = \{\}) \\ & where\ \rho' = \rho \cup tags(\texttt{timer}^{\ell}(\hat{x}(y).\mathbf{0},Q)) \\ & and\ \rho'' = \rho \cup tags(\texttt{timer}^{\ell}(\hat{x}(y).P,\mathbf{0})) \end{array}\right.$

$(\gamma_{time}6) \quad \gamma_{time}([\![\texttt{timer}^{t^{\ell}}(\hat{x}().P,Q)]\!])\rho =$
$\left\{\begin{array}{ll} \texttt{timer}^{\ell}(\hat{x}().\gamma_{time}([\![P]\!])\rho',Q) & if\ tags(\gamma_{time}([\![P]\!])\rho')\backslash tags(P) \neq \{\} \\ \texttt{timer}^{\ell}(\hat{x}().P,\gamma_{time}([\![Q]\!])\rho'') & if\ (tags(\gamma_{time}([\![P]\!])\rho')\backslash tags(P) = \{\}) \wedge \\ & \quad (tags(\gamma_{time}([\![Q]\!])\rho'')\backslash tags(Q) \neq \{\}) \\ \texttt{timer}^{\ell}(\hat{x}().P,Q) & if\ (tags(\gamma_{time}([\![P]\!])\rho')\backslash tags(P) = \{\}) \wedge \\ & \quad (tags(\gamma_{time}([\![Q]\!])\rho'')\backslash tags(Q) = \{\}) \\ & where\ \rho' = \rho \cup tags(\texttt{timer}^{\ell}(\hat{x}().\mathbf{0},Q)) \\ & and\ \rho'' = \rho \cup tags(\texttt{timer}^{\ell}(\hat{x}().P,\mathbf{0})) \end{array}\right.$

$(\gamma_{time}7) \quad \gamma_{time}([\![P \mid Q]\!])\rho \qquad\quad =$
$\left\{\begin{array}{ll} \gamma_{time}([\![P]\!])\rho' \mid Q & if\ tags(\gamma_{time}([\![P]\!])\rho')\backslash tags(P) \neq \{\} \\ P \mid \gamma_{time}([\![Q]\!])\rho'' & if\ (tags(\gamma_{time}([\![P]\!])\rho')\backslash tags(P) = \{\}) \wedge \\ & \quad (tags(\gamma_{time}([\![Q]\!])\rho'')\backslash tags(Q) \neq \{\}) \\ P \mid Q & if\ (tags(\gamma_{time}([\![P]\!])\rho')\backslash tags(P) = \{\}) \wedge \\ & \quad (tags(\gamma_{time}([\![Q]\!])\rho'')\backslash tags(Q) = \{\}) \\ & where\ \rho' = \rho \cup tags(Q) \\ & and\ \rho'' = \rho \cup tags(P) \end{array}\right.$

$(\gamma_{time}8) \quad \gamma_{time}([\![!P]\!])\rho \qquad\qquad\quad = !\gamma_{time}([\![P]\!])\rho$

$(\gamma_{time}9) \quad \gamma_{time}([\![(\nu x)P]\!])\rho \qquad\quad = (\nu x)\gamma_{time}([\![P]\!])\rho$

$(\gamma_{time}10) \quad \gamma_{time}([\![\mathbf{0}]\!])\rho \qquad\qquad\quad = \mathbf{0}$

$(\gamma_{time}11) \quad \gamma_{time}([\![P+Q]\!])\rho \qquad\quad =$
$\left\{\begin{array}{ll} \gamma_{time}([\![P]\!])\rho' + Q & if\ tags(\gamma_{time}([\![P]\!])\rho')\backslash tags(P) \neq \{\} \\ P + \gamma_{time}([\![Q]\!])\rho'' & if\ (tags(\gamma_{time}([\![P]\!])\rho')\backslash tags(P) = \{\}) \wedge \\ & \quad (tags(\gamma_{time}([\![Q]\!])\rho'')\backslash tags(Q) \neq \{\}) \\ P + Q & if\ (tags(\gamma_{time}([\![P]\!])\rho')\backslash tags(P) = \{\}) \wedge \\ & \quad (tags(\gamma_{time}([\![Q]\!])\rho'')\backslash tags(Q) = \{\}) \\ & where\ \rho' = \rho \cup tags(Q) \\ & and\ \rho'' = \rho \cup tags(P) \end{array}\right.$

Fig. 6.4 Definition of the time-tagging function

$$tags(R) = \begin{cases} tags(P) & \text{if } R = \overline{x}\langle y\rangle.P \ \vee \ R = \overline{x}\langle\rangle.P \ \vee \\ & \quad R = !P \ \vee \ R = (\nu x)P \\ \{\ell\} \cup tags(P) & \text{if } R = \overline{x}^{\ell}\langle y\rangle.P \ \vee \ R = \overline{x}\langle y^{\ell}\rangle.P \ \vee \\ & \quad R = \overline{x}^{\ell}\langle\rangle.P \\ \{\ell, \ell'\} \cup tags(P) & \text{if } R = \overline{x}^{\ell}\langle y^{\ell'}\rangle.P \\ \{\ell, \ell'\} \cup tags(P) \cup tags(Q) & \text{if } R = \texttt{timer}^{t^{\ell}}(x^{\ell'}(y).P, Q) \ \vee \\ & \quad R = \texttt{timer}^{t^{\ell}}(x^{\ell'}().P, Q) \\ \{\ell\} \cup tags(P) \cup tags(Q) & \text{if } R = \texttt{timer}^{t^{\ell}}(x(y).P, Q) \ \vee \\ & \quad R = \texttt{timer}^{t^{\ell}}(x().P, Q) \ \vee \\ & \quad R = \texttt{timer}^{t}(x^{\ell}(y).P, Q) \ \vee \\ & \quad R = \texttt{timer}^{t}(x^{\ell}().P, Q) \\ tags(P) \cup tags(Q) & \text{if } R = \texttt{timer}^{t}(x(y).P, Q) \ \vee \\ & \quad R = \texttt{timer}^{t}(x().P, Q) \ \vee \\ & \quad R = P \mid Q \ \vee \ R = P + Q \\ \{\} & \text{if } R = \mathbf{0} \end{cases}$$

In Fig. 6.2, we tag the channel names of a process in a syntax-directed manner.

The environment, $\rho \subseteq \mathcal{T}$, holds the set of all the tags that the tagging function has so far "seen". The rules can be described as follows. Rules $(\gamma_{chn}1)$ and $(\gamma_{chn}2)$ tag an untagged channel name for the non-empty and empty output actions, respectively, using a tag that has not yet been seen before by the tagging function and does not belong to the set of tags of the process itself. Rules $(\gamma_{chn}3)$ and $(\gamma_{chn}4)$ pass the tagging function to the body of a non-empty and empty output action, respectively, where the channel of the action is a tagged one while recording that tag (as well as any tag the output message may carry) in ρ. In rules $(\gamma_{chn}5)$ and $(\gamma_{chn}6)$, a non-empty and empty timed input, respectively, which has an untagged channel, is tagged. This is done again in a way that ensures that the chosen tag is not one of the existing ones in ρ or in the timed input itself. In rules $(\gamma_{chn}7)$ and $(\gamma_{chn}8)$, a non-empty and empty timed input, respectively, using a tagged channel name, will have the tagging function passed to either of the two continuation processes of the action, or none. This will depend on whether we find other channel names that we can tag in either of the two processes, or none. If we did, then the difference between the newly tagged process and the original one will be a non-empty set. In rules $(\gamma_{chn}9)$ and $(\gamma_{chn}13)$, we use the same idea to tag either side of a parallel composition and a non-deterministic choice, or neither, if neither can be tagged any further. In rules $(\gamma_{chn}10)$ and $(\gamma_{chn}11)$, we pass the tagging function to the process inside the operator enclosing it. Finally, in rule $(\gamma_{chn}12)$, the channel tagging function has no effect on a null process.

By contrast, Fig. 6.3 outlines the rules of the message-tagging function.

In these rules, rule $(\gamma_{msg}1)$ tags an untagged message name in a non-empty output action, using a tag that has not yet been seen by the tagging function and does not belong to the set of tags of the process itself. Rule $(\gamma_{msg}2)$ passes the tagging function to the body of a non-empty output action with a tagged message while recording that tag (as well as any tag the channel name may carry) in ρ. In rule $(\gamma_{msg}3)$, the same is done for the case of an empty output action, since such action would have no message to tag. In rules $(\gamma_{msg}4)$ and $(\gamma_{msg}5)$, non-empty and empty timed input actions naturally have no output messages, therefore, we pass the tagging function to either of the two continuation processes, or none, depending on whether we find other channel names that we can tag in either of these two continuation processes (or none). If we did, then the difference between the newly tagged process and the original one will be a non-empty set. In rules $(\gamma_{msg}6)$ and $(\gamma_{msg}10)$, we use the same idea to tag either side of a parallel composition and a non-deterministic choice, or none if neither have message names that can be tagged. In rules $(\gamma_{msg}7)$ and $(\gamma_{msg}8)$, we pass the message-tagging function to the process inside the operator enclosing it. Finally, in rule $(\gamma_{msg}9)$, the message-tagging function has no effect on a null process.

Finally, Fig. 6.4 shows the rules of the input time-tagging function.

The rules can be described as follows. Rules $(\gamma_{time}1)$ and $(\gamma_{time}2)$ do not tag anything in non-empty and empty output actions, respectively, and simply pass the tagging function to the body of the output action, while recording any tags the action may carry in ρ. Rules $(\gamma_{time}3)$ $(\gamma_{time}4)$ tag an untagged time value of a non-empty or empty input action, respectively, using a tag that has not been recorded in ρ so far and does not belong to the tags of the input process itself. In rules $(\gamma_{time}5)$ and $(\gamma_{time}6)$, we pass the tagging function to either of the two continuation processes or neither, for a non-empty or empty input action, respectively, where the time value has already been tagged. In rules $(\gamma_{time}7)$ and $(\gamma_{time}11)$, we use the same idea to tag either side of a parallel composition and a non-deterministic choice, or none if neither have time values that can be tagged. In rules $(\gamma_{time}8)$ and $(\gamma_{time}9)$, we pass the time-tagging function to the process inside the operator enclosing it. Finally, in rule $(\gamma_{time}7)$, the time-tagging function has no effect on a null process.

We can prove the termination of the tagging functions, as follows.

Theorem 6.1 (Termination of the Tagging Functions) *The aforementioned γ_{chn}, γ_{msg} and γ_{time} tagging functions terminate.*

Proof sketch. This is possible to prove by showing that the size-change principle applies [8]. First, we define a size function on the number of terms that may need to be tagged:

$$
size(R) = \begin{cases}
1 + size(P) & \text{if } R = \overline{\hat{x}}\langle\hat{y}\rangle.P \\
1 + size(P) + size(Q) & \text{if } R = \texttt{timer}^f(\hat{x}(y).P, Q) \\
size(P) & \text{if } R = !P \ \vee \ R = (\nu x)P \\
size(P) + size(Q) & \text{if } R = P \mid Q \ \vee \ R = P + Q \\
0 & \text{if } R = \mathbf{0}
\end{cases}
$$

We can express any of the rules for γ_{chn}, γ_{msg} and γ_{time} in one of the following two generic forms:

$$\gamma(\![R]\!)\rho = C[\gamma(\![P]\!)\rho']$$
$$\gamma(\![R]\!)\rho = P$$

where $C[.]$ is the context of a process, defined itself as being a process with a hole. For the first form, it is possible to show by structural induction on the rules of γ_{chn}, γ_{msg} and γ_{time} that it is always the case that

$size(P) < size(R)$

This proves that the size-change principle applies here. For the second form, the tagging function naturally comes to a termination point. $\square$

Theorem 6.2 (Completeness) *Every relevant non-tagged term is tagged.*

Proof sketch. The proof of this lemma falls from the proof of Theorem 6.1 and by structural induction on the rules of the tagging functions, γ_{chn}, γ_{msg} and γ_{time}, which show that no non-tagged terms are missed. $\square$

6.4.2 Application of the Tagging Functions

The application of the tagging functions depends on what combination of mutations we are seeking to generate, and this will depend further on the scenario the mutation analysis is being used for. However, the most general case is that we would like to tag every channel name, message and time value. In order to do that, we first define the following function:

$$\omega = \lambda f.\lambda \gamma.\lambda p.\lambda \rho.(if \gamma (\![p]\!)\rho = p \ then \ p \ else \ f \ \gamma \ (\gamma (\![p]\!)\rho) \ tags(\gamma (\![p]\!)\rho)) \quad (6.5)$$

then for a specific tagging function, γ, we compute its fixed point using Curry's fixed-point combinator [9, p. 178]:

$$\mathbf{Y} \ \omega \ \gamma \ P \ tags(P) = \omega \ (\mathbf{Y} \ \omega) \ \gamma \ P \ tags(P) \quad (6.6)$$

If we name the above fixed point for the γ tagging function, $fix_\gamma(P)$, then the full application of our three tagging functions to a process, P, yielding our final process, P_{final}, can be expressed as follows:

$$P_{final} = fix_{\gamma_{time}} (fix_{\gamma_{msg}} (fix_{\gamma_{chn}} (P)))$$

6.4.3 Definition of the General Mutation Function

We now define a general mutation function, in Fig. 6.5, that takes a tagged process and a specific tag, and produces a mutant, which is another (possibly tagged) process:

$$\mu : \mathcal{P}^\mathcal{T} \times \mathcal{T} \times ((\mathcal{N} \cup \mathbb{N}) \to (\mathcal{N} \cup \mathbb{N})) \to \mathcal{P}^\mathcal{T} \quad (6.7)$$

In these rules, the function $f : (\mathcal{N} \cup \mathbb{N}) \to (\mathcal{N} \cup \mathbb{N})$ is a polymorphic name- and time-changing function, defined as follows:

$$(\mu 1) \ \mu([\bar{\hat{x}}\langle \hat{y} \rangle.P])\ell f \quad = \begin{cases} \overline{z}\langle \hat{y} \rangle.P & \text{if } \hat{x} = x^{\ell}, \text{ where } z = f(x) \\ \bar{\hat{x}}\langle z \rangle.P & \text{if } \hat{y} = y^{\ell}, \text{ where } z = f(y) \\ \bar{\hat{x}}\langle \hat{y} \rangle.\mu([P])\ell f & \text{otherwise} \end{cases}$$

$$(\mu 2) \ \mu([\bar{\hat{x}}\langle \rangle.P])\ell f \quad = \begin{cases} \overline{z}\langle \rangle.P & \text{if } \hat{x} = x^{\ell}, \text{ where } z = f(x) \\ \bar{\hat{x}}\langle \rangle.\mu([P])\ell f & \text{otherwise} \end{cases}$$

$(\mu 3) \ \mu([\texttt{timer}^{\hat{t}}(\hat{x}(y).P, Q)])\ell f =$
$$\begin{cases} \texttt{timer}^{t'}(\hat{x}(y).P, Q) & \text{if } \hat{t} = t^{\ell}, \text{ where } t' = f(t) \\ \texttt{timer}^{\hat{t}}(z(y).P, Q) & \text{if } \hat{x} = x^{\ell}, \text{ where } z = f(x) \\ \texttt{timer}^{\hat{t}}(\hat{x}(y).\mu([P])\ell f, Q) & \text{if } P \neq \mu([P])\ell f \\ \texttt{timer}^{\hat{t}}(\hat{x}(y).P, \mu([Q])\ell f) & \text{otherwise} \end{cases}$$

$(\mu 4) \ \mu([\texttt{timer}^{\hat{t}}(\hat{x}().P, Q)])\ell f \ =$
$$\begin{cases} \texttt{timer}^{t'}(\hat{x}().P, Q) & \text{if } \hat{t} = t^{\ell}, \text{ where } t' = f(t) \\ \texttt{timer}^{\hat{t}}(z().P, Q) & \text{if } \hat{x} = x^{\ell}, \text{ where } z = f(x) \\ \texttt{timer}^{\hat{t}}(\hat{x}().\mu([P])\ell f, Q) & \text{if } P \neq \mu([P])\ell f \\ \texttt{timer}^{\hat{t}}(\hat{x}().P, \mu([Q])\ell f) & \text{otherwise} \end{cases}$$

$$(\mu 5) \ \mu([P \mid Q])\ell f \quad = \begin{cases} \mu([P])\ell f \mid Q & \text{if } P \neq \mu([P])\ell f \\ P \mid \mu([Q])\ell f & \text{otherwise} \end{cases}$$

$$(\mu 6) \ \mu([!P])\ell f \quad = \ !\mu([P])\ell f$$

$$(\mu 7) \ \mu([(\nu x)P])\ell f \quad = \ (\nu x)\mu([P])\ell f$$

$$(\mu 8) \ \mu([\mathbf{0}])\ell f \quad = \ \mathbf{0}$$

$$(\mu 9) \ \mu([P + Q])\ell f \quad = \begin{cases} \mu([P])\ell + Q & \text{if } P \neq \mu([P])\ell f \\ P + \mu([Q])\ell f & \text{otherwise} \end{cases}$$

Fig. 6.5 Definition of the general mutation function

$$f(r) : \begin{cases} g(r) \ \text{if } r \in \mathcal{N} \\ h(r) \ \text{if } r \in \mathbb{N} \end{cases} \tag{6.8}$$

where $g : \mathcal{N} \to \mathcal{N}$ is a name-changing function and $h : \mathbb{N} \to \mathbb{N}$ is a time-changing function. The actual definitions of g and h depend on the specific scenario. For example, g could be defined as the most obvious function that selects randomly some name in $\mathcal{N}$ not equal to the name it receives:

$$g(x) = y, \ \text{where } y \in \mathcal{N} \text{ and } y \neq x$$

The case of h is more interesting. This is because time, in the context of the timed input action, may have different effects. Take, for example, $\texttt{timer}^{\infty}(x(y).P, Q)$ which means that the input action will always be available (and hence Q is practically non-reachable). On the other hand, $\texttt{timer}^{0}(x(y).P, Q)$ means that the input action itself is bypassed and Q is the only practically available process. Therefore, the choice of how we mutate t is important. Here, we are interested in choices of h that cover three important points on the time line: the initial point ($t = 0$), a random midpoint ($0 < t < \infty$) and the point of infinity ($t = \infty$), since these three points affect the behaviour of input actions at the abstract level. Therefore, we consider the following three definitions of h (called $h_{\to 0}$, $h_{\to \infty}$ and $h_{\to t}$):

$$h_{\to 0}(t) = 0$$
$$h_{\to \infty}(t) = \infty$$
$$h_{\to t}(t) = \begin{cases} t' & \text{if } t = 0 \ \vee \ t = \infty, \text{ where } 0 < t' < \infty \\ t & \text{otherwise} \end{cases}$$

$h_{\to 0}$ changes every time value to the initial point in time. $h_{\to \infty}$, on the other hand, changes every time value to the infinite point in time. And finally, $h_{\to t}$ changes both the two extremes to some middle point in time, or if the point is already somewhere in the middle, it returns the value unchanged. There are other definitions that h can be given, e.g. defining h as a *piece-wise* function. That will depend on the context of the system being analysed. However, in our case, we don't think that our static analysis later benefits from such other definitions.

The set of all mutants that can be derived from a process, P, is

$$\{m \mid (m = \mu(\![P]\!)\ell f) \ \wedge \ (\ell \in tags(P)) \ \wedge \ (f \in F)\}$$

where F is the space of all name- and time-changing functions, f, specifically adopted in the case of the mutation scenario. To ensure that the mutation of a process terminates, $|F|$ must be finite.

6.4.4 Mutating the OCPP Heartbeat Protocol

6.4.4.1 Charging Station Mutants

The first set of mutants were generated for the Charging Station process. There are 16 such mutants, resulting from alterations to channel names (mutants $m1$–$m7$), output messages (mutants $m8$–$m10$) and input time values (mutants $m11$–$m16$). The definitions of all of these mutants are shown in Figs. 6.6 and 6.7.

6.4.4.2 CSMS Mutants

The second group of mutants are the CSMS mutants. There are 9 such mutants, resulting from alterations to channel names (mutants $m17$–$m20$), output message

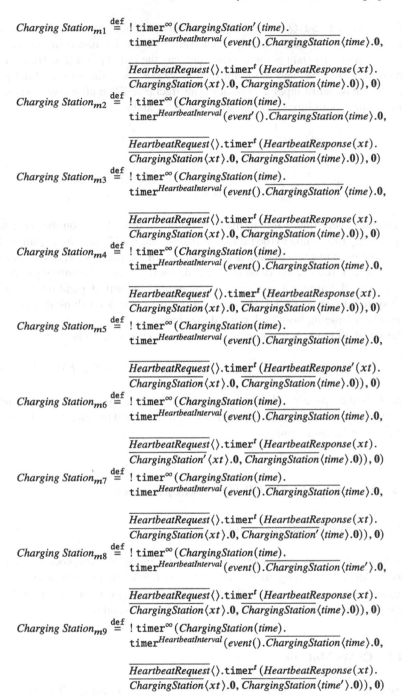

$Charging\ Station_{m1} \stackrel{\text{def}}{=}\ !\ \text{timer}^\infty\,(ChargingStation'\,(time).$
$\qquad\qquad\qquad \text{timer}^{HeartbeatInterval}\,(event().\overline{ChargingStation}\,\langle time\rangle.\mathbf{0},$

$\qquad\qquad\qquad \overline{HeartbeatRequest}\,\langle\rangle.\text{timer}^t\,(HeartbeatResponse(xt).$
$\qquad\qquad\qquad \overline{ChargingStation}\,\langle xt\rangle.\mathbf{0},\overline{ChargingStation}\,\langle time\rangle.\mathbf{0})),\mathbf{0})$

$Charging\ Station_{m2} \stackrel{\text{def}}{=}\ !\ \text{timer}^\infty\,(ChargingStation(time).$
$\qquad\qquad\qquad \text{timer}^{HeartbeatInterval}\,(event'().\overline{ChargingStation}\,\langle time\rangle.\mathbf{0},$

$\qquad\qquad\qquad \overline{HeartbeatRequest}\,\langle\rangle.\text{timer}^t\,(HeartbeatResponse(xt).$
$\qquad\qquad\qquad \overline{ChargingStation}\,\langle xt\rangle.\mathbf{0},\overline{ChargingStation}\,\langle time\rangle.\mathbf{0})),\mathbf{0})$

$Charging\ Station_{m3} \stackrel{\text{def}}{=}\ !\ \text{timer}^\infty\,(ChargingStation(time).$
$\qquad\qquad\qquad \text{timer}^{HeartbeatInterval}\,(event().\overline{ChargingStation'}\,\langle time\rangle.\mathbf{0},$

$\qquad\qquad\qquad \overline{HeartbeatRequest}\,\langle\rangle.\text{timer}^t\,(HeartbeatResponse(xt).$
$\qquad\qquad\qquad \overline{ChargingStation}\,\langle xt\rangle.\mathbf{0},\overline{ChargingStation}\,\langle time\rangle.\mathbf{0})),\mathbf{0})$

$Charging\ Station_{m4} \stackrel{\text{def}}{=}\ !\ \text{timer}^\infty\,(ChargingStation(time).$
$\qquad\qquad\qquad \text{timer}^{HeartbeatInterval}\,(event().\overline{ChargingStation}\,\langle time\rangle.\mathbf{0},$

$\qquad\qquad\qquad \overline{HeartbeatRequest'}\,\langle\rangle.\text{timer}^t\,(HeartbeatResponse(xt).$
$\qquad\qquad\qquad \overline{ChargingStation}\,\langle xt\rangle.\mathbf{0},\overline{ChargingStation}\,\langle time\rangle.\mathbf{0})),\mathbf{0})$

$Charging\ Station_{m5} \stackrel{\text{def}}{=}\ !\ \text{timer}^\infty\,(ChargingStation(time).$
$\qquad\qquad\qquad \text{timer}^{HeartbeatInterval}\,(event().\overline{ChargingStation}\,\langle time\rangle.\mathbf{0},$

$\qquad\qquad\qquad \overline{HeartbeatRequest}\,\langle\rangle.\text{timer}^t\,(HeartbeatResponse'(xt).$
$\qquad\qquad\qquad \overline{ChargingStation}\,\langle xt\rangle.\mathbf{0},\overline{ChargingStation}\,\langle time\rangle.\mathbf{0})),\mathbf{0})$

$Charging\ Station_{m6} \stackrel{\text{def}}{=}\ !\ \text{timer}^\infty\,(ChargingStation(time).$
$\qquad\qquad\qquad \text{timer}^{HeartbeatInterval}\,(event().\overline{ChargingStation}\,\langle time\rangle.\mathbf{0},$

$\qquad\qquad\qquad \overline{HeartbeatRequest}\,\langle\rangle.\text{timer}^t\,(HeartbeatResponse(xt).$
$\qquad\qquad\qquad \overline{ChargingStation'}\,\langle xt\rangle.\mathbf{0},\overline{ChargingStation}\,\langle time\rangle.\mathbf{0})),\mathbf{0})$

$Charging\ Station_{m7} \stackrel{\text{def}}{=}\ !\ \text{timer}^\infty\,(ChargingStation(time).$
$\qquad\qquad\qquad \text{timer}^{HeartbeatInterval}\,(event().\overline{ChargingStation}\,\langle time\rangle.\mathbf{0},$

$\qquad\qquad\qquad \overline{HeartbeatRequest}\,\langle\rangle.\text{timer}^t\,(HeartbeatResponse(xt).$
$\qquad\qquad\qquad \overline{ChargingStation}\,\langle xt\rangle.\mathbf{0},\overline{ChargingStation'}\,\langle time\rangle.\mathbf{0})),\mathbf{0})$

$Charging\ Station_{m8} \stackrel{\text{def}}{=}\ !\ \text{timer}^\infty\,(ChargingStation(time).$
$\qquad\qquad\qquad \text{timer}^{HeartbeatInterval}\,(event().\overline{ChargingStation}\,\langle time'\rangle.\mathbf{0},$

$\qquad\qquad\qquad \overline{HeartbeatRequest}\,\langle\rangle.\text{timer}^t\,(HeartbeatResponse(xt).$
$\qquad\qquad\qquad \overline{ChargingStation}\,\langle xt\rangle.\mathbf{0},\overline{ChargingStation}\,\langle time\rangle.\mathbf{0})),\mathbf{0})$

$Charging\ Station_{m9} \stackrel{\text{def}}{=}\ !\ \text{timer}^\infty\,(ChargingStation(time).$
$\qquad\qquad\qquad \text{timer}^{HeartbeatInterval}\,(event().\overline{ChargingStation}\,\langle time\rangle.\mathbf{0},$

$\qquad\qquad\qquad \overline{HeartbeatRequest}\,\langle\rangle.\text{timer}^t\,(HeartbeatResponse(xt).$
$\qquad\qquad\qquad \overline{ChargingStation}\,\langle xt\rangle.\mathbf{0},\overline{ChargingStation}\,\langle time'\rangle.\mathbf{0})),\mathbf{0})$

Fig. 6.6 Charging station mutants

$Charging\ Station_{m10} \overset{\text{def}}{=}\ !\ \text{timer}^{\infty}\,(ChargingStation(time).$
$\qquad \text{timer}^{HeartbeatInterval}\,(\overline{event()}.ChargingStation\langle time\rangle.0,$

$\qquad \overline{HeartbeatRequest\langle\rangle}.\text{timer}^{t}\,(HeartbeatResponse(xt).$
$\qquad ChargingStation\langle xt'\rangle.0,\ \overline{ChargingStation\langle time\rangle.0)),0)$

$Charging\ Station_{m11} \overset{\text{def}}{=}\ !\ \text{timer}^{t}\,(ChargingStation(time).$
$\qquad \text{timer}^{HeartbeatInterval}\,(\overline{event()}.ChargingStation\langle time\rangle.0,$

$\qquad \overline{HeartbeatRequest\langle\rangle}.\text{timer}^{t}\,(HeartbeatResponse(xt).$
$\qquad ChargingStation\langle xt\rangle.\mathbf{0},\ \overline{ChargingStation\langle time\rangle.0)),0)$

$Charging\ Station_{m12} \overset{\text{def}}{=}\ !\ \text{timer}^{0}\,(ChargingStation(time).$
$\qquad \text{timer}^{HeartbeatInterval}\,(\overline{event()}.ChargingStation\langle time\rangle.0,$

$\qquad \overline{HeartbeatRequest\langle\rangle}.\text{timer}^{t}\,(HeartbeatResponse(xt).$
$\qquad ChargingStation\langle xt\rangle.0,\ \overline{ChargingStation\langle time\rangle.0)),0)$

$Charging\ Station_{m13} \overset{\text{def}}{=}\ !\ \text{timer}^{\infty}\,(ChargingStation(time).$
$\qquad \text{timer}^{0}\,(\overline{event()}.ChargingStation\langle time\rangle.0,$

$\qquad \overline{HeartbeatRequest\langle\rangle}.\text{timer}^{t}\,(HeartbeatResponse(xt).$
$\qquad ChargingStation\langle xt\rangle.0,\ \overline{ChargingStation\langle time\rangle.0)),0)$

$Charging\ Station_{m14} \overset{\text{def}}{=}\ !\ \text{timer}^{\infty}\,(ChargingStation(time).$
$\qquad \text{timer}^{\infty}\,(\overline{event()}.ChargingStation\langle time\rangle.0,$

$\qquad \overline{HeartbeatRequest\langle\rangle}.\text{timer}^{t}\,(HeartbeatResponse(xt).$
$\qquad ChargingStation\langle xt\rangle.0,\ \overline{ChargingStation\langle time\rangle.0)),0)$

$Charging\ Station_{m15} \overset{\text{def}}{=}\ !\ \text{timer}^{\infty}\,(ChargingStation(time).$
$\qquad \text{timer}^{HeartbeatInterval}\,(\overline{event()}.ChargingStation\langle time\rangle.0,$

$\qquad \overline{HeartbeatRequest\langle\rangle}.\text{timer}^{0}\,(HeartbeatResponse(xt).$
$\qquad ChargingStation\langle xt\rangle.0,\ \overline{ChargingStation\langle time\rangle.0)),0)$

$Charging\ Station_{m16} \overset{\text{def}}{=}\ !\ \text{timer}^{\infty}\,(ChargingStation(time).$
$\qquad \text{timer}^{HeartbeatInterval}\,(\overline{event()}.ChargingStation\langle time\rangle.0,$

$\qquad \overline{HeartbeatRequest\langle\rangle}.\text{timer}^{\infty}\,(HeartbeatResponse(xt).$
$\qquad ChargingStation\langle xt\rangle.0,\ \overline{ChargingStation\langle time\rangle.0)),0)$

Fig. 6.7 Charging station mutants (cont.)

names (mutant $m21$) and input time values (mutants $m22$–$m25$). These mutants are shown in Fig. 6.8.

$$CSMS_{m17} \stackrel{\text{def}}{=} \; ! \, \texttt{timer}^{\infty}(CSMSc'().\texttt{timer}^{\infty}(HeartbeatRequest().$$

$$\overline{HeartbeatResponse}\langle currentTime\rangle.\overline{CSMSc}\langle\rangle.\mathbf{0}, \mathbf{0}), \mathbf{0})$$

$$CSMS_{m18} \stackrel{\text{def}}{=} \; ! \, \texttt{timer}^{\infty}(CSMSc().\texttt{timer}^{\infty}(HeartbeatRequest'().$$

$$\overline{HeartbeatResponse}\langle currentTime\rangle.\overline{CSMSc}\langle\rangle.\mathbf{0}, \mathbf{0}), \mathbf{0})$$

$$CSMS_{m19} \stackrel{\text{def}}{=} \; ! \, \texttt{timer}^{\infty}(CSMSc().\texttt{timer}^{\infty}(HeartbeatRequest().$$

$$\overline{HeartbeatResponse'}\langle currentTime\rangle.\overline{CSMSc}\langle\rangle.\mathbf{0}, \mathbf{0}), \mathbf{0})$$

$$CSMS_{m20} \stackrel{\text{def}}{=} \; ! \, \texttt{timer}^{\infty}(CSMSc().\texttt{timer}^{\infty}(HeartbeatRequest().$$

$$\overline{HeartbeatResponse}\langle currentTime\rangle.\overline{CSMSc'}\langle\rangle.\mathbf{0}, \mathbf{0}), \mathbf{0})$$

$$CSMS_{m21} \stackrel{\text{def}}{=} \; ! \, \texttt{timer}^{\infty}(CSMSc().\texttt{timer}^{\infty}(HeartbeatRequest().$$

$$\overline{HeartbeatResponse}\langle currentTime'\rangle.\overline{CSMSc}\langle\rangle.\mathbf{0}, \mathbf{0}), \mathbf{0})$$

$$CSMS_{m22} \stackrel{\text{def}}{=} \; ! \, \texttt{timer}^{t}(CSMSc().\texttt{timer}^{\infty}(HeartbeatRequest().$$

$$\overline{HeartbeatResponse}\langle currentTime\rangle.\overline{CSMSc}\langle\rangle.\mathbf{0}, \mathbf{0}), \mathbf{0})$$

$$CSMS_{m23} \stackrel{\text{def}}{=} \; ! \, \texttt{timer}^{0}(CSMSc().\texttt{timer}^{\infty}(HeartbeatRequest().$$

$$\overline{HeartbeatResponse}\langle currentTime\rangle.\overline{CSMSc}\langle\rangle.\mathbf{0}, \mathbf{0}), \mathbf{0})$$

$$CSMS_{m24} \stackrel{\text{def}}{=} \; ! \, \texttt{timer}^{\infty}(CSMSc().\texttt{timer}^{t}(HeartbeatRequest().$$

$$\overline{HeartbeatResponse}\langle currentTime\rangle.\overline{CSMSc}\langle\rangle.\mathbf{0}, \mathbf{0}), \mathbf{0})$$

$$CSMS_{m25} \stackrel{\text{def}}{=} \; ! \, \texttt{timer}^{\infty}(CSMSc().\texttt{timer}^{0}(HeartbeatRequest().$$

$$\overline{HeartbeatResponse}\langle currentTime\rangle.\overline{CSMSc}\langle\rangle.\mathbf{0}, \mathbf{0}), \mathbf{0})$$

Fig. 6.8 CSMS mutants

$Heartbeat_{m26} \overset{\text{def}}{=} (\nu ChargingStation)(\nu CSMSc)($
$\qquad Charging\ Station\ |\ \overline{ChargingStation'}\langle zerotime\rangle.0\ |\ CSMS\ |$
$\qquad \overline{CSMSc}\langle\rangle.0)$

$Heartbeat_{m27} \overset{\text{def}}{=} (\nu ChargingStation)(\nu CSMSc)($
$\qquad Charging\ Station\ |\ \overline{ChargingStation}\langle zerotime\rangle.0\ |\ CSMS\ |$
$\qquad \overline{CSMSc'}\langle\rangle.0)$

$Heartbeat_{m28} \overset{\text{def}}{=} (\nu ChargingStation)(\nu CSMSc)($
$\qquad Charging\ Station\ |\ \overline{ChargingStation}\langle zerotime'\rangle.0\ |\ CSMS\ |$
$\qquad \overline{CSMSc}\langle\rangle.0)$

Fig. 6.9 Protocol-level mutants

6.4.4.3 Protocol-Level Mutants

The final group of mutations were generated at the protocol level itself. There are 3 mutants at this level, covering changes to channel names (mutants $(m26–m27)$) and output messages (mutant $m28$). These mutants are shown in Fig. 6.9. There are no time value mutations as there are no input actions at the protocol's level.

6.4.5 Analysing the Mutants

To understand later the impact mutations will have on the protocol, we start here by analysing the Heartbeat protocol itself without any mutations applied. Therefore, we perform the analysis, $\mathcal{A}(\![Heartbeat]\!)\phi_0 = \phi_{normal}$, where $\phi_{normal} = \phi' \cup_\phi \phi''$, and where

$\phi'[time_1 \mapsto zeroTime, time_2 \mapsto time_1, time_2 \mapsto time_2]$
$\phi''[time_1 \mapsto zeroTime, xt_1 \mapsto currentTime, time_2 \mapsto xt_1, xt_2 \mapsto currentTime,$
$\quad time_2 \mapsto xt_2]$

We analysed each case of the Heartbeat protocol mutants $(m1–m28)$ under an approximation that places a bound on the number of copies of input parameters and freshly generated names. We consider a bound of *two* such copies only. Results of the analysis are shown in Tables 6.1 and 6.2.

Each case represents one change, in the Charging Station process, the CSMS process or in the protocol process itself. We have analysed each run of a mutant, $Heartbeat_{mi}$ (for $i = 1\ldots28$), in parallel with two events emitted from the CSMS:

$$\mathcal{A}(\![Heartbeat_{mi}\ |\ \overline{event}\langle\rangle.0\ |\ \overline{event}\langle\rangle.0]\!)\phi_0$$

Table 6.1 Results of the static analysis of the heartbeat protocol

Mutant	Analysis result
$Heartbeat_{m1}$	ϕ_0
$Heartbeat_{m2}$	$\phi[time_1 \mapsto zeroTime, xt_1 \mapsto currentTime, time_2 \mapsto xt_1,$ $xt_2 \mapsto currentTime, time_2 \mapsto xt_2]$
$Heartbeat_{m3}$	$\phi = \phi' \cup_\phi \phi''$ $where, \phi'[time_1 \mapsto zeroTime] \ and$ $\phi''[time_1 \mapsto zeroTime, xt_1 \mapsto currentTime, time_2 \mapsto xt_1,$ $xt_2 \mapsto currentTime, time_2 \mapsto xt_2]$
$Heartbeat_{m4}$	$\phi = \phi' \cup_\phi \phi_0$ $where, \phi'[time_1 \mapsto zeroTime, time_2 \mapsto time_1, time_2 \mapsto time_2]$
$Heartbeat_{m5}$	$\phi = \phi' \cup_\phi \phi''$ $where, \phi'[time_1 \mapsto zeroTime, time_2 \mapsto time_1, time_2 \mapsto time_2] \ and$ $\phi''[time_1 \mapsto zeroTime, time_2 \mapsto time_1, time_2 \mapsto time_2]$
$Heartbeat_{m6}$	$\phi = \phi' \cup_\phi \phi''$ $where, \phi'[time_1 \mapsto zeroTime, time_2 \mapsto time_1, time_2 \mapsto time_2] \ and$ $\phi''[time_1 \mapsto zeroTime, xt_1 \mapsto currentTime, xt_2 \mapsto currentTime]$
$Heartbeat_{m7}$	$\phi = \phi' \cup_\phi \phi''$ $where, \phi'[time_1 \mapsto zeroTime, time_2 \mapsto time_1, time_2 \mapsto time_2] \ and$ $\phi''[time_1 \mapsto zeroTime, xt_1 \mapsto currentTime, time_2 \mapsto xt_1,$ $xt_2 \mapsto currentTime, time_2 \mapsto xt_2]$
$Heartbeat_{m8}$	$\phi = \phi' \cup_\phi \phi''$ $where, \phi'[time_1 \mapsto zeroTime, time_2 \mapsto time'] \ and$ $\phi''[time_1 \mapsto zeroTime, xt_1 \mapsto currentTime, time_2 \mapsto xt_1,$ $xt_2 \mapsto currentTime, time_2 \mapsto xt_2]$
$Heartbeat_{m9}$	$\phi = \phi' \cup_\phi \phi''$ $where, \phi'[time_1 \mapsto zeroTime, time_2 \mapsto time_1, time_2 \mapsto time_2] \ and$ $\phi''[time_1 \mapsto zeroTime, xt_1 \mapsto currentTime, time_2 \mapsto xt_1,$ $xt_2 \mapsto currentTime, time_2 \mapsto xt_2]$
$Heartbeat_{m10}$	$\phi = \phi' \cup_\phi \phi''$ $where, \phi'[time_1 \mapsto zeroTime, time_2 \mapsto time_1, time_2 \mapsto time_2] \ and$ $\phi''[time_1 \mapsto zeroTime, xt_1 \mapsto currentTime, time_2 \mapsto xt',$ $xt_2 \mapsto currentTime]$
$Heartbeat_{m11}$	$\phi = \phi' \cup_\phi \phi''$ $where, \phi'[time_1 \mapsto zeroTime, time_2 \mapsto time_1, time_2 \mapsto time_2] \ and$ $\phi''[time_1 \mapsto zeroTime, xt_1 \mapsto currentTime, time_2 \mapsto xt_1,$ $xt_2 \mapsto currentTime, time_2 \mapsto xt_2]$
$Heartbeat_{m12}$	ϕ_0

Table 6.2   Results of the static analysis of the heartbeat protocol (cont.)

Mutant	Analysis result
$Heartbeat_{m}13$	$\phi[time_1 \mapsto zeroTime, xt_1 \mapsto currentTime, time_2 \mapsto xt_1,$
	$\quad xt_2 \mapsto currentTime, time_2 \mapsto xt_2]$
$Heartbeat_{m}14$	$\phi[time_1 \mapsto zeroTime, time_2 \mapsto time_1, time_2 \mapsto time_2]$
$Heartbeat_{m}15$	$\phi = \phi' \cup_\phi \phi''$
	where, $\phi'[time_1 \mapsto zeroTime, time_2 \mapsto time_1, time_2 \mapsto time_2]$ and
	$\phi''[time_1 \mapsto zeroTime, time_2 \mapsto time_1, time_2 \mapsto time_2]$
$Heartbeat_{m}16$	$\phi = \phi' \cup_\phi \phi''$
	where, $\phi'[time_1 \mapsto zeroTime, time_2 \mapsto time_1, time_2 \mapsto time_2]$ and
	$\phi''[time_1 \mapsto zeroTime, xt_1 \mapsto currentTime, time_2 \mapsto xt_1,$
	$\quad xt_2 \mapsto currentTime, time_2 \mapsto xt_2]$
$Heartbeat_{m}17$	$\phi[time_1 \mapsto zeroTime, time_2 \mapsto time_1, time_2 \mapsto time_2]$
$Heartbeat_{m}18$	$\phi[time_1 \mapsto zeroTime, time_2 \mapsto time_1, time_2 \mapsto time_2]$
$Heartbeat_{m}19$	$\phi[time_1 \mapsto zeroTime, time_2 \mapsto time_1, time_2 \mapsto time_2]$
$Heartbeat_{m}20$	$\phi = \phi' \cup_\phi \phi''$
	where, $\phi'[time_1 \mapsto zeroTime, time_2 \mapsto time_1, time_2 \mapsto time_2]$ and
	$\phi''[time_1 \mapsto zeroTime, xt_1 \mapsto currentTime, time_2 \mapsto xt_1]$
$Heartbeat_{m}21$	$\phi = \phi' \cup_\phi \phi''$
	where, $\phi'[time_1 \mapsto zeroTime, time_2 \mapsto time_1, time_2 \mapsto time_2]$ and
	$\phi''[time_1 \mapsto zeroTime, xt_1 \mapsto currentTime', time_2 \mapsto xt_1,$
	$\quad xt_2 \mapsto currentTime', time_2 \mapsto xt_2]$
$Heartbeat_{m}22$	$\phi = \phi' \cup_\phi \phi''$
	where, $\phi'[time_1 \mapsto zeroTime, time_2 \mapsto time_1, time_2 \mapsto time_2]$ and
	$\phi''[time_1 \mapsto zeroTime, xt_1 \mapsto currentTime, time_2 \mapsto xt_1,$
	$\quad xt_2 \mapsto currentTime, time_2 \mapsto xt_2]$
$Heartbeat_{m}23$	$\phi[time_1 \mapsto zeroTime, time_2 \mapsto time_1, time_2 \mapsto time_2]$
$Heartbeat_{m}24$	$\phi = \phi' \cup_\phi \phi''$
	where, $\phi'[time_1 \mapsto zeroTime, time_2 \mapsto time_1, time_2 \mapsto time_2]$ and
	$\phi''[time_1 \mapsto zeroTime, xt_1 \mapsto currentTime, time_2 \mapsto xt_1,$
	$\quad xt_2 \mapsto currentTime, time_2 \mapsto xt_2]$
$Heartbeat_{m}25$	$\phi[time_1 \mapsto zeroTime, time_2 \mapsto time_1, time_2 \mapsto time_2]$
$Heartbeat_{m}26$	ϕ_0
$Heartbeat_{m}27$	$\phi[time_1 \mapsto zeroTime, time_2 \mapsto time_1, time_2 \mapsto time_2]$
$Heartbeat_{m}28$	$\phi = \phi' \cup_\phi \phi''$
	where, $\phi'[time_1 \mapsto zeroTime', time_2 \mapsto time_1, time_2 \mapsto time_2]$ and
	$\phi''[time_1 \mapsto zeroTime', xt_1 \mapsto currentTime, time_2 \mapsto xt_1,$
	$\quad xt_2 \mapsto currentTime, time_2 \mapsto xt_2]$

Table 6.3 Mutant classes according to the analysis results

Class	Indistinguishability set
Normal Case	$\{Heartbeart_{m7}, Heartbeart_{m9}, Heartbeart_{m11}, Heartbeart_{m16},$
	$Heartbeart_{m22}, Heartbeart_{m24}\}$
Inactive	$\{Heartbeart_{m1}, Heartbeart_{m12}, Heartbeart_{m26}\}$
No-heartbeat	$\{Heartbeart_{m4}, Heartbeart_{m14}, Heartbeart_{m17}, Heartbeart_{m18},$
	$\{Heartbeart_{m19}, Heartbeart_{m23}, Heartbeart_{m25}, Heartbeart_{m27}\}$
Different Update Time	$\{Heartbeart_{m21}\}$
Different Initial Time	$\{Heartbeart_{m28}\}$
Class 1	$\{Heartbeart_{m5}, Heartbeart_{m15}\}$
Class 2	$\{Heartbeart_{m2}, Heartbeart_{m3}, Heartbeart_{m13}\}$
Class 3	$\{Heartbeart_{m6}\}$
Class 4	$\{Heartbeart_{m8}\}$
Class 5	$\{Heartbeart_{m10}\}$
Class 6	$\{Heartbeart_{m20}\}$

where these events (referred to as CSMS activity in the informal description of the scenario in [1, p. 185]) allow the Charging Station to follow the normal case where no heartbeat is required. Our choice of two events only coincides with the approximation bound on the number of copies mentioned above, as more events would have generated indistinguishable copies. We also limit our choice of the name- and time-changing function space to the following:

$$F = \{(\lambda g, h.f) \; g \; h_{\to 0}, (\lambda g, h.f) \; g \; h_{\to \infty}, (\lambda g, h.f) \; g \; h_{\to t}\}$$

The results of the analysis reveal the following classes of indistinguishable mutants, shown in Table 6.3. Note that our analysis is purely testing the effects of mutations in an otherwise non-malicious setup. In other words, we do not consider the presence of any adversaries. Such an adversarial environment would require a different, more detailed, analysis.

We now discuss each of these mutation classes, at an informal level.

Normal Class

This case represents all the mutants that have analysis results similar to results obtained in the normal (un-mutated) case, and therefore, they are indistinguishable from the normal specification. In other words, $\phi = \phi_{normal}$. These mutants are harmless, in the context of our analysis, as they do not affect the behaviour of the system.

Inactive Class

This class represents all the cases where the mutation caused the complete stalling of the protocol. As a result, $\phi = \phi_0$. These mutants cause "deadly" failures, and hence, they are easy to spot.

No-heartbeat Class
This class represents all the mutants that exhibit only behaviour corresponding to responses to the CSMS external events, over *event*(), but that are never able to enter the heartbeat stage if no such external events are received. These mutants represent partial behaviour according to what the standard protocol requires.

Different Update Time Class
This mutant represents the case where the updating of the time of the Charging Station, at the end of the heartbeat stage, is done using a different value than the current time at the CSMS. This mutant is an example of harmful mutations, since a wrong current time value can have further effects on the overall OCPP protocol. For example, it can influence the time validity of the "ChargingProfile" datatype used by the CSMS to influence, through the Charging Station, the power or current drawn by an EV (see Use Case K01 [1, p. 233]).

Different Initial Time Class
This mutant represents the case where the Charging Station is initialised with a time value different from the normal initial time, when it bootstraps (see Use Case BO1 [1, p. 43]). Again, if the Charging station is used by an EV prior to its first heartbeat communication with the CSMS, this could cause a negative impact on the charging profile, as stated for the mutant above.

Classes 1–6
These classes of mutants represent various effects that have no high-level interesting impact on the system, other than changing its operational behaviour. Therefore, we do not discuss them any further.

It is worth noting here that the above eleven classes we identified are *specific* to the type of analysis we discussed at the beginning of this section, i.e. the name-substitution analysis. Our method is more general in that any other type of analysis can equally be applied leading to a (possibly) different classification of mutants with different effects and impacts on the protocol's behaviour and its properties. One of the interesting future ideas would be to apply data-driven (quantitative) analysis methods to understand if changes in system behaviour lead to changes measurable to the environment within which the system is operating. For example, how changes in numbers of messages or changes in time ranges would affect performance.

References

1. OCPP 2.0.1 Part 2 - Specification. https://www.openchargealliance.org/protocols/ocpp-201/. Accessed 23 Apr 2020
2. M.C. Falvo, D. Sbordone, I.S. Bayram, M. Devetsikiotis, in *2014 International Symposium on Power Electronics, Electrical Drives, Automation and Motion* (IEEE, 2014), pp. 1134–1139
3. S.S. Hussain, T.S. Ustun, P. Nsonga, I. Ali, IEEE Trans. Veh. Technol. **67**(8), 7690 (2018)
4. D.J. Andrews, T.L. Polmateer, J.P. Wheeler, D.L. Slutzky, J.H. Lambert, Curr. Sustain./Renew. Energy Rep. pp. 1–7 (2020)
5. Y. Jia, M. Harman, IEEE Trans. Software Eng. **37**(5), 649 (2010)

6. Open Charge Alliance. https://www.openchargealliance.org/. Accessed 11 May 2020
7. OCPP 2.0.1 Part 1 - Architecture & Topology. https://www.openchargealliance.org/protocols/ocpp-201/. Accessed 23 Apr 2020
8. C.S. Lee, N.D. Jones, A.M. Ben-Amram, SIGPLAN Not. **36**(3), 8192 (2001)
9. H.B. Curry, R. Feys, *Combinatory Logic*, vol. 1 (North-Holland Publishing Company, Amsterdam, 1958). https://books.google.co.uk/books?id=fEnuAAAAMAAJ

Chapter 7
Conclusion

7.1 Introduction

The modern advances in the design and implementation of distributed and mobile computational systems and the advent of IoT, Industry 4.0 and smart infrastructures have increased the demands for the understanding of the behaviour of such systems in a robust and rigorous manner in order to better protect and run such systems with increased efficiency with as few errors as possible. For example, the task of preventing unauthorised malicious intruders and erroneous programs from compromising the secrecy, authenticity and other desirable security properties is growing into a complex and delicate task as a result of the dynamic nature of the network topologies such modern systems and protocols are characterised by. This dynamic nature facilitated a plethora of novel methods by which attacks from such intruders could be mounted.

One significant approach to the analysis of program behaviour at compile-time is static analysis, which is used to determine properties about programs and their expected runtime behaviour prior to their execution. This is necessary to determine whether programs exhibit certain security flaws, like information leakage and lack of authenticity, as well as whether they adhere to their standard specifications. It is also necessary when designing newer systems and languages to avoid ambiguity and increase the coverage of testing suites. Therefore, static analysers constitute standard tools often used as part of language compilers. Their applications reach beyond the areas of code safety and security into other areas like program optimisation and transformation.

In this concluding chapter, we summarise the work presented in each chapter of the book providing prospects for future research.

109
B. Aziz, *Formal Analysis by Abstract Interpretation*, SpringerBriefs in Applied Sciences and Technology, https://doi.org/10.1007/978-3-030-91153-9_7

7.2   Process Algebra

This chapter presented the syntax and structural operational semantics of the formal language used throughout the book, which is a form of a process algebra based on the π-calculus enhanced with timed input constructs. The chapter also defines a non-standard semantics for this language, aimed at capturing a fundamental property in communicating systems, namely message passing leading to name substitutions. This non-standard semantics allows for the interpretation to reflect how various input parameters in a process are instantiated. We proved the soundness of the non-standard denotational semantics with respect to the standard structural operational semantics of the language.

7.3   Formal Analysis by Abstract Interpretation

This chapter defined a method for approximating the non-standard semantics capturing message passing defined in Chap. 2. The approximation is based on limiting the number of distinguishable copies a process can produce of new names and input parameters. This then leads to the generation of a finite semantic domain and an abstract interpretation function that is used to give an abstract semantics for a process, which is then proved to be safe with respect to the concrete non-standard semantics. The chapter was concluded with a couple of simple examples to demonstrate how the analysis can be used in understanding the behaviour of systems. These examples included a simple FTP system and a simple distance-bounding protocol system. In the FTP example, we demonstrated that using per-session passwords (once-off passwords) protects authenticity, even though the passwords were leaked after their usage. In the distance-bounding protocol, the results of the analysis were used to define a name integrity property and a notion of MitM attacks under a definition for the *fastest* Dolev–Yao intruder.

7.4   First Case Study

This chapter modelled and analysed, as the first case study, the MQ Telemetry Transport protocol, which is a lightweight broker-based publish–subscribe protocol that is used in communications with small devices that exhibit limited computational and storage power. We found that the first two QoS modes of operation in the protocol are clearly specified, and their message delivery semantics to subscribers can be easily verified to hold. However, according to the results of the analysis, the last case of an "exactly once" delivery semantics has potential vulnerabilities where a simple attacker model that adheres to the specified threat model of the protocol can cause the semantics to be undermined. At best, this semantics is vaguely specified in the standard [1], particularly in relation to issues to do with the choice of server-side behaviour.

As directions of future research, one could focus on studying the properties of the protocol under more aggressive attacker models and we plan to propose refined versions of the protocol, including the use of lightweight cryptography in scenarios where the authentication of the small devices is required. In addition, although we carried out a simple modification to the QoS = 2 case that removes the duplicated publish message vulnerability, we would like to further investigate in-depth additional mechanisms for improving further the protocol. This would call for more automated approaches, namely using any of a number of automated verification tools that exist in the literature, e.g. [2–7].

7.5 Second Case Study

This chapter demonstrated how our static analysis can be used in combination with a testing framework to consolidate some parts of the Hermes protocol in its original specification document [8], as the second case study in our book. The analysis showed that the protocol, despite the lack of consideration for simultaneous errors, behaves in an intended manner. We used a formal specification and analysis approach to demonstrate that upstream and downstream machine agreement in every case was as expected according to the protocol's specification, where the transfer completes successfully, due to a lack of errors, or does not start or complete, due to the presence of errors at some stage of the board transfer process.

Future direction of work here would include modelling and analysing other Industry 4.0 communication technologies [9], e.g. the Open Platform Communications (OPC) Unified Architecture [10], Bosch's Production Performance Management Protocol (PPMP) [11] and the Data Distribution Service [12]. We also plan to specify probabilistic and stochastic properties and verify whether such properties hold in the Hermes protocol and other Industry 4.0 protocols. Another interesting area to extend this research to would be the new IPC-CFX (Connected Factory Exchange) standard [13], which transforms a Hermes-based manufacturing line into a fully Industrial Internet-of-Things environment. As far as we know, the CFX standard is not yet made open-source.

7.6 Third Case Study

This chapter defined, as the third case study, a formal mutation framework that can be used to introduce mutants of systems and protocols specified using a variation of the π-calculus. The framework uses a single general mutation function, which we used to perform mutations on messages, names of communication channels and duration of input actions. The mutation function was then applied to a case study of an electric vehicle charging protocol standard, called OCPP. Results of the analysis revealed eleven classes of mutants; some were indistinguishable from the normal un-mutated case, some had no impact on the system and some had interesting implications for the safety and security of the protocol.

Future directions for research could include combining the mutation function μ with other types of analysis, $\mathcal{A}$, which captures other behaviour different from the name substitutions, and hence, we can observe other effects of mutations created by μ. Another interesting area would be to extend the definition of the mutation function itself to include other mutations of process specifications. More specifically, these could include changes to the structure of the process itself, for example, in the manner in which subprocesses are combined whether in parallel or a non-deterministic choice. There is also, of course, the prospect of applying the current mutation and analysis functions to a different process algebra, or indeed, a different formal language altogether, if one wanted to cover some other class of systems. Finally, although this chapter did not consider any models of adversaries since the main concern was to demonstrate only the effects of mutations on systems, however, an immediate line of research more concerned with the security and safety of systems would be to consider adding the adversary to the analysis of the OCPP protocol (or for that matter, any other protocol under similar study). Such an adversary (e.g. Dolev–Yao's most powerful adversary [14]) can easily be modelled as a process running in parallel with the specific mutant being analysed.

References

1. D. Locke, MQ Telemetry Transport (MQTT) V3.1 Protocol Specification. Technical report, IBM Corporation (2010)
2. ProVerif: Cryptographic protocol verifier in the formal model. http://prosecco.gforge.inria.fr/personal/bblanche/proverif/. Accessed 24 Sep 2014
3. D.A. Basin, S. Mödersheim, L. Viganò, Int. J. Inf. Sec. 4(3), 181 (2005)
4. The Tamarin prover for security protocol analysis. https://hackage.haskell.org/package/tamarin-prover. Accessed 24 Sep 2014
5. Maude-NPA. http://maude.cs.uiuc.edu/tools/Maude-NPA/. Accessed 24 Sep 2014
6. Casper: A Compiler for the Analysis of Security Protocols. http://www.cs.ox.ac.uk/gavin.lowe/Security/Casper/. Accessed 24 Sep 2014
7. The Scyther Tool. http://www.cs.ox.ac.uk/people/cas.cremers/scyther/index.html. Accessed 24 Sep 2014
8. T.H.S. Initiative; IPC-HERMES-9852: The global standard for machine-to-machine communication in SMT assembly (version 1.2). Technical report, IPC (2019)
9. P. Marcon, F. Zezulka, I. Vesely, Z. Szabo, Z. Roubal, O. Sajdl, E. Gescheidtova, P. Dohnal, in *2017 Progress In Electromagnetics Research Symposium - Spring (PIERS)* (2017), pp. 1694–1697
10. The OPC Unified Architecture. https://opcfoundation.org/about/opc-technologies/opc-ua/. Accessed 11 Jun 2019
11. The PPMP Specification. https://www.eclipse.org/unide/specification/. Accessed 11 Jun 2019

12. The Data Distribution Service Foundation. https://www.dds-foundation.org/. Accessed 11 Jun 2019
13. C.F. Initiative, IPC-2591: Connected Factory Exchange (CFX). Technical report, IPC (2019)
14. D. Dolev, A. Yao, in *Proceedings of the 22nd Annual Symposium on Foundations of Computer Science* (1981), pp. 350–357

Printed in the United States
by Baker & Taylor Publisher Services